Towards a Dynamic Regional Innovation System

Wenying Fu

Towards a Dynamic Regional Innovation System

Investigation into the Electronics Industry in the Pearl River Delta, China

Wenying Fu
School of Geographical Science
South China Normal University
Guangzhou
Guangdong
China

ISBN 978-3-662-45415-2   ISBN 978-3-662-45416-9 (eBook)
DOI 10.1007/978-3-662-45416-9

Library of Congress Control Number: 2014954803

Springer Heidelberg New York Dordrecht London
© Springer-Verlag Berlin Heidelberg 2015
This work is subject to copyright. All rights are reserved by the Publisher, whether the whole or part of the material is concerned, specifically the rights of translation, reprinting, reuse of illustrations, recitation, broadcasting, reproduction on microfilms or in any other physical way, and transmission or information storage and retrieval, electronic adaptation, computer software, or by similar or dissimilar methodology now known or hereafter developed.
The use of general descriptive names, registered names, trademarks, service marks, etc. in this publication does not imply, even in the absence of a specific statement, that such names are exempt from the relevant protective laws and regulations and therefore free for general use.
The publisher, the authors and the editors are safe to assume that the advice and information in this book are believed to be true and accurate at the date of publication. Neither the publisher nor the authors or the editors give a warranty, express or implied, with respect to the material contained herein or for any errors or omissions that may have been made.

Printed on acid-free paper

Springer is part of Springer Science+Business Media (www.springer.com)

Acknowledgement

The completion of this study witnessed my growth as an independent researcher during my overseas study at Institute of Economic and Cultural Geography, Leibniz University of Hannover, Germany. As a doctoral dissertation project, the work has been supported by the German Research Foundation (DFG) under framework of the Priority Program 1233 "*Megacity-Megachallenge: Informal Dynamics of Global Change*". The financial support it has offered has made the field survey and data collection possible. In addition, the organization committee of the cooperated project has devoted a lot of efforts into organizing regular interdisciplinary workshops and conferences, as well as supporting my international exchange activities, all of which have provided me precious opportunities in learning the latest theoretical and methodological development in the field of innovation studies. In the following-up studies after the DFG program, acknowledgement should be made to the Project "*Study on industrial cluster restructuring mechanism from the perspective of local entrepreneurship: Case study in the Pearl River Delta, China*" (No. 41301109) supported by the Natural Science Foundation of China.

I am very much indebted to a great number of persons in the process of the study. Foremost, I want to say thank you in all sincerity to my supervisor Prof. Dr. Javier Revilla Diez. He constantly challenged me to refine the theoretical discussions and to explore deeper for the implications of the empirical results. Owing to his German-style instruction, famous for the tedious training of logical reasoning, this work has been greatly improved. Javier is very supportive for my career development after I graduated, both financially and professionally. I would also like to acknowledge the co-examiner of the work, Prof. Dr. Ingo Liefner, especially for his generous knowledge spillover in the workshop discussions.

Special thanks need to go to Dr. Susanne Meyer and PD Dr. Daniel Schiller for their efforts in making my research and life in Hannover a valuable experience. Susanne has taken great care of me since the first day I arrived in Germany, and she has set up a role model for me with her outstanding research outcomes in the first phase of the Priority Program. Daniel has also devoted a lot of time in reading through my far-from-perfect drafts and giving me very constructive advices to improve the work. Furthermore, three key members in the "agile firm" research project under the DFG program, PD Dr. Stefan Hennemann, Dr. Wan-Hsin Liu and

Dr. Stephan Ohm, have played important role in the successful completion of the survey project, and deserve special acknowledgement for the numerous fruitful and illuminating discussions that we've had together. Without your joint efforts, the research project would not have been such a success. I would also like to mention Prof. Lundvall, Prof. Kerndrup, Prof. Keun Lee and Buru Im, who have given useful suggestions for my presentation, as part of the study, in the DRUID Winter and Summer Conference 2011.

Last but not least, I would like to dedicate this work to my family, who have supported me unconditionally and always love me just as sure as the stars shine above.

Contents

1 **Introduction** 1
   1.1 Research Context 1
   1.2 Defining Innovation and Regional Innovation System
       in China Context 3
       1.2.1 What does Innovation Imply in China? 3
       1.2.2 What does Regional Innovation System Imply in China? 5
   1.3 Aims and Research Questions 6
   1.4 Outline 8
   1.5 Survey Data and Evaluation 10
   References 14

2 **Knowledge Spillovers and Regional Innovation: The Case of
Guangdong Province, China** 17
   2.1 Introduction 17
   2.2 Technological Upgrading and Innovation in Guangdong
       Province, China: Some Stylized Facts 20
   2.3 Impact of Knowledge Spillover on Innovation: An
       Integrative Perspective from the Local and the Global Scale 24
   2.4 Model Specification and Data 30
       2.4.1 Data and Measurements 31
       2.4.2 Static and Dynamic Approach for Panel Data 33
       2.4.3 Description of the Data 34
   2.5 Empirical Results 36
   2.6 Discussion and Conclusion 39
   References 41

3 **Interactive Learning and Systemic Innovation** 45
   3.1 Introduction 45
   3.2 Innovation as an Interactive Process 47
   3.3 Survey Data and Methodology 52
   3.4 Empirical Results 53
       3.4.1 Descriptive Results 53
       3.4.2 Econometrical Analysis 56

	3.5	Discussion and Conclusion	63
	References		64
4	**Absorptive Capacity, Proximity and Innovation: Linking up the Intra-Firm Characteristic with Inter-Firm Linkages**		**67**
	4.1	Introduction	68
	4.2	Use of Proximity in Interactive Learning	70
		4.2.1 Proximity: Concept and Taxonomy	70
		4.2.2 Organizational Proximity and Social Proximity: Comparison and Dynamics	72
		4.2.3 Proximity for the SMEs in the Clusters	79
		4.2.4 Brief Summary	81
	4.3	Absorptive Capacity in the Firm Level as Precondition of Interactive Learning	83
		4.3.1 Human Capital	84
		4.3.2 R&D Activities	85
		4.3.3 Production Experience	87
		4.3.4 Brief Summary	88
	4.4	Operationalization of Analysis	89
	4.5	Empirical Evidence	91
		4.5.1 Innovation Behavior of Electronics Firms	91
		4.5.2 Absorptive Capacity and Learning Behaviors	100
		4.5.3 The SMEs' Use of Proximity	107
		4.5.4 Impact of the Use of Proximity on Product Innovation Outcome	109
	4.6	Discussion and Conclusion	115
	References		117
5	**From Globalized Production Systems to Regional Innovation Systems: Governance and Innovation in Shenzhen and Dongguan, China**		**123**
	5.1	Introduction	123
	5.2	Evolutionary Regional Innovation System and Governance Infrastructure	125
		5.2.1 Evolution of Governance Infrastructure: Content and Typology	125
		5.2.2 Evolution of Governance Infrastructure: Dynamics and Inertia	129
	5.3	Survey Design of a Comparative Study	131
	5.4	Governance in Shenzhen and Dongguan, China: An Evolutionary Overview	133
		5.4.1 Governance Evolution in Shenzhen Since Opening	133
		5.4.2 Governance Evolution in Dongguan Since Opening	136
		5.4.3 Summary of Governance in Shenzhen and Dongguan	138

	5.5		Descriptive Profile of Innovation Activities in Shenzhen and Dongguan	139
	5.6		Empirical Evidence for Interactive Innovation	142
	5.7		Discussion and Conclusion	150
	References			151
6	**Conclusions**			155
	6.1		What do We Learn About the Chinese Regional Innovation System?	155
		6.1.1	The Shaping of Interactive Learning Behavior and Systemic Innovation	156
		6.1.2	The Informal Aspect of Innovation Activities in China	157
		6.1.3	The Spatial Difference under Divergent Governance Infrastructure Evolution	159
	6.2		Directions of Future Research	160
		6.2.1	The Mechanism of Distributive System on the Regional Level	160
		6.2.2	The Negative Effect of Informal Guanxi Network on Innovation	161
		6.2.3	Methodological Issues in the Survey Design	161
	6.3		Policy Implications	162
		6.3.1	Enhancing and Balancing the Firm Absorptive Capacity	163
		6.3.2	Identifying and Supporting the Capacity of Interactive Learning	163
		6.3.3	Timely Assessing the Inertia Governance Infrastructure	164
	References			165
7	**Appendices**			167
	7.1		Appendix A: Firm Questionnaire	167
	7.2		Appendix B: Test of Clusterin Solution	180
	7.3		Appendix C: Classifying Product Technology	181
	7.4		Appendix D: Development of Shenzhen Electronics Industry in 1980s and 1990s	185

List of Abbreviation

AIC	Akaike information criteria
BIC	Bayesian information criteria
CEO	Chief executive officer
CPI	Consumer price index
DFG	German research foundation
FDI	Foreign direct investment
GDP	Gross domestic product
GECC	Guangdong electronic chamber of commerce
HHI	Hirschman-Herfindahl index
LCD	Liquid crystal display
MAR	Marshall-Arrow-Romer
NPCK	New product codified knowledge
NPI	New product ideas
NPTK	New product tacit knowledge
OECD	Organization for economic co-operation and development
ODM	Original design manufacturer
OEM	Original equipment manufacturer
OBM	Original brand manufacturer
OLS	Ordinary least squares
PC	Personal computer
PPI	Production price index
PRD	Pearl River Delta
R&D	Research and development
RIS	Regional innovation system
RMB	Renminbi (Chinese currency)
SCPRC	State council of the People's Republic of China
SECC	Shenzhen electronic chamber of commerce
SME	Small and medium-sized enterprise
TFP	Total factor productivity

Chapter 1
Introduction

Abstract This chapter serves as an introductory note to the research context, key concepts and primary database upon which the book develops. Situated within the context of economic restructuring around the globe, China's high-speed growth has been curtailed by both internal cost pressure and external market change. Under such circumstances, technological upgrading and innovation capabilities is the key to the successful restructuring process. The work adopts the regional innovation system approach, which proposes the institutional and organizational dimension as the supporting infrastructure that stabilizes the interactive learning process. By investigating the electronics industry in the Pearl River Delta (PRD), one of the biggest megacity regions in China, this work aims to address three primary questions. Firstly, it will explore theoretically and empirically the external channels that are able to trigger the local-scale knowledge spillover. Secondly, it aims to expand the understanding of the role of informality in reducing uncertainties and risk faced with innovation activities. Last but not least, the spatial differences and processes in regard to firm innovation is analyzed.

1.1 Research Context

Driven by economic liberalization and globalization, Chinese has grown as an economic powerhouse since the opening-up policy in the late 1970s. Yet in recent years, its high-speed growth has been substantially constrained by both internal and external factors in recent years. On one hand, the high inflation rate that leads to continual pressure of rising costs gradually erodes the competitive edge on low cost production. In the first quarter of 2011, the Consumer Price Index (CPI) hit the record of 5.4% year-on-year, and the Production Price Index (PPI) also followed the rising trend, reaching 7.3% for the first quarter[1]. On the other hand, Chinese export firms are encountered with more trade obstacles in the developed market due to the protection of local employment market after the financial crisis. Firms either have to meet the high standards on safety and quality in order to maintain the market share in developed countries, or they have to exploit the new market opportunities in the domestic economy.

---
[1] The data is posted by China Statistical Bureau, April 2011.

In this circumstance, technological upgrading and innovation capabilities is the key to the successful restructuring process. The innovation investment cools down the fervent economic growth owing to its long period of returning rate, and at the same time ensure the sustainable growth engine in the long run. In response to the call of the innovation issue within the context of inflationary growth and competition pressure, China's innovation policy has been greatly focused on science & technology policy (SCPRC 2006), aiming to foster indigenous innovation capabilities through Research and Development (R&D) incentivized tax reduction, improving intellectual property rights and setting its own technological standards. In other words, the Chinese innovation policy follows a linear legacy, in which innovation is taken as a sequential process of discovery and direct translation into commercial value.

Nevertheless, this linear approach underestimates the interactive and systemic nature of innovation in value creation (Lundvall 1992; Cooke et al.1997; Howells 1999; Revilla Diez 2000; Smith 2000; Asheim and Coenen 2005). The system approach towards innovation has been proposed in the innovation milieu by Aydalot (1986), in cluster theory by Porter (1990), in national innovation systems by Lundvall (1992) and in regional innovation systems by Cooke et al. (1997), all of whom have recognized the interactive learning process and the resulting distribution power of a production system as the fundamental element of economic performance. In this way, the knowledge exploitation process in the economy yields increasing returns on the generated knowledge, propelling the endogenous process of economic growth.

As a latecomer country, China has the advantage of backwardness, in which the technological knowledge is available "off the shelf" (Nolan and Lenski 1985). Consequently, knowledge exploitation is more important than knowledge generation. For latecomers, access to technology in industrialized countries as well as successful absorption and translation into market opportunities, combined with the low-cost and flexible manufacturing advantage, constitute the core elements of their competitiveness. Therefore, innovation potential in China can be at best released by implementing effective technology transfer and strengthening the distributive power of the economic system as a whole.

The distributive power of the system depends on the willingness and capability of local firms to undertake interactive learning. The regional innovation system approach proposes the institutional and organizational dimension as the supporting infrastructure that stabilizes the interactive learning process. Heidenreich (2004) defines the stabilizing factor as the regional orders, encompassing formalized rules and laws as well as informal habits and methods. The regional orders promote the interactive learning process and systemic innovation activities by reducing uncertainty, coordinating the use of knowledge and mediating conflicts.

Overall, this book aims to explore the formation process and characteristics of the regional innovation system in China, which is of great relevance to the release of innovation potential in the face of upgrading pressure. As demonstrated by Heidenreich (2004), the strength of a regional innovation system does not lie in the static set of institutions, firms and technologies, but in its dynamic ability to overcome dilemmas and meet the challenge of market change and organizational restructuring. Therefore, the dynamic and evolutionary perspective on the regional innovation system is adopted in this book so that signs of a maturing regional innovation system can be captured, investigated and compared with regard to both the business superstructure and the governance infrastructure.

The research is supported within the framework of the Priority Programme 1233 "Megacity-Megachallenge: Informal Dynamics of Global Change" funded by the German Research Foundation (DFG). In this research program, one of the biggest megacity regions in China, the Pearl River Delta (PRD), has been selected as the research region. The electronics industry forms the particular focus of the study.

The electronics industry has been developing in the PRD for over 30 years. For strategic reasons, nearly 90% of the global lead firms in the electronics industry have located themselves in the east coastal cities of the PRD in particular, such as Shenzhen and Dongguan. The electronics industry in this region is very export-oriented. The region manufactures over 50% of the world's desktop computers and 40% of Personal Computer (PC) components, such as PC heads, PC cases and other semi-manufactured products[2]. Moreover, many domestic brands in the PRD have rapidly developed and taken a considerable share of the global market. However, with the increasing land and labor costs in the PRD, and the favorable policies offered by many inland governments—further buttressed by the regional bias towards central and western provinces with the 4 trillion stimulus package after the 2008 crisis, the trend of industrial shift to inland China is irreversible. Therefore, the externally-driven growth mode is no longer sustainable, and there is an urgent call for the development of regional innovation system to generate sustainable and dynamic growth paths.

Moreover, the electronics industry has a large pool of technological opportunities, which confers the great possibility of opening up numerous niche markets with new product development. Firms can profit in niche markets by minor innovation when prerequisite absorptive capability, such as the ability to understand and adjust the circuit board design, is ready. For minor innovators in electronics industry, interactive learning with users and other knowledge-intensive organizations assists in collecting market information and supported technology.

> One of the Shenzhen exhibitors in "China Sourcing Fair: Electronics & Components" displayed their new product—Solar Charger Backpack. The manager told the journalist that the orders have reached over 10 million Yuan. "What we do is just to make the collection and the use of solar energy more convenient, but this minor innovation led to higher added value for our products."
> —Shenzhen News, 04.2011

1.2 Defining Innovation and Regional Innovation System in China Context

1.2.1 What does Innovation Imply in China?

As China is a technological latecomer, the content of innovation is more incremental than abrupt. Knowledge production activities are not dominant in these

---
[2] Source: http://www.gdiid.gd.gov.cn/gdiid/billion/lay2-3.htm. Accessed 15th September 2014.

countries, since the modern natural sciences, such as physics, chemistry, biology and so on, are led by industrialized countries. R&D activity, which is a main proxy of knowledge production activities, displays an unbalanced pattern between industrialized countries and latecomer countries. Although R&D expenditure has greatly increased, e.g. to 1.7% of Gross Domestic Product (GDP) in 2009, the intensity is still fairly weak compared to that of developed countries (Organization for Economic Co-operation and Development, hereafter shortened for OECD 2.3% in 2009, USA 2.9% in 2009, Japan 3.4% in 2009, Korea 3.3% in 2009; See OECD 2011). Therefore, access to advanced knowledge and dissemination mechanisms remains the key factor for successful incremental innovation in the Chinese context (OECD 2005).

Overall, innovation in China is characterized as:

1. Resource restriction of firm-level innovation. Because of the low entry barriers to simple assembly processing tasks, small and medium-sized firms are dominant in latecomer countries. The lack of economies of scale leads to resistance to conducting high risk innovation activities at the firm level. Furthermore, the underdeveloped financial system provides weak financial support for firms to invest in innovation.
2. Unbalanced knowledge base and weaker regional innovation system. The industrialization process only began in China some 30 years ago. The industrial knowledge base is weak and unevenly distributed among firms in the region. As a result, the mismatch of absorptive capability among firms in the region can hardly generate knowledge spillover to stimulate the cross-fertilization among the firms. Furthermore, the linkages between universities, research institutes and business firms are fairly weak.
3. Reliance on external sources for innovation. The globalization process is transforming from vertical disintegration within a lead company to organizational fragmentation, which spreads more widely into low-cost regions, and thus exerting hierarchical network control on the upgrading and innovation of firms in China. Codified technology transfer embedded in import goods as well as codified and tacit technology transfer from multinational corporations is, therefore, a fundamental source of innovation.
4. Unstable institutional system. In transition China, the market mechanism is not fully developed and the institutional environment is undergoing a continual transition and reform process. In this context, firms face unexpected costs and risks which inhibit them from engaging in long-term innovation activities. Moreover, local protectionism undermines the economies of scale and expected innovation return, which reduces to a certain degree the incentive of firms to innovate.
5. Informality. In the uncertain environment in China, firms tend to apply an informal network-based strategy. The informal relations among firms, which are mostly sustained through *Guanxi* networks with relatives, friends and business partners, have contributed to the flexible and responsive production which has further strengthened China's low-cost manufacturing strategy. Provided a maturing and balanced development of the absorptive capacity for Chinese firms,

informal networking is likely to play a more important role than the formal institutional framework in constructing regional orders that facilitate the distribution and exploitation of external advanced knowledge.

1.2.2 What does Regional Innovation System Imply in China?

The term 'regional innovation system' is widely understood as "interacting knowledge generation and exploitation sub-systems for commercializing new knowledge" (Cooke 2004, p. 3). Braczyk et al. (1998) proposed a two-dimensional structure for understanding the function of this territorial sub-system, consisting of the governance infrastructure and the business superstructure. The governance infrastructure supports the competitiveness of firms' business performance and linkages towards each other and with the outside world, and stabilizes the interacting process of knowledge generation and exploitation with established orders, encompassing physical organizations such as research competence, education, funding and technological transfer agencies as well as socio-institutional rules and norms.

As indicated in the above discussion on the characteristics of innovation in China, the regional innovation system is still weak due to the scarcity of innovation-related resources, capabilities and institutions. As the production activities are highly dependent on the foreign direct investment (FDI) in the developed coastal regions of China, the prospect of a well-functioning regional innovation system lies in its capacity to capitalize on the external linkages for commercializing new knowledge. Figure 1.1 graphically demonstrates the implication of a well-functioning regional innovation system within this context in China. It consists primarily of two general aspects: the exploitation by firms of both the external knowledge (mainly from foreign investment) and the local interdependency for enhancing the competitiveness (Asheim and Isaksen 2002).

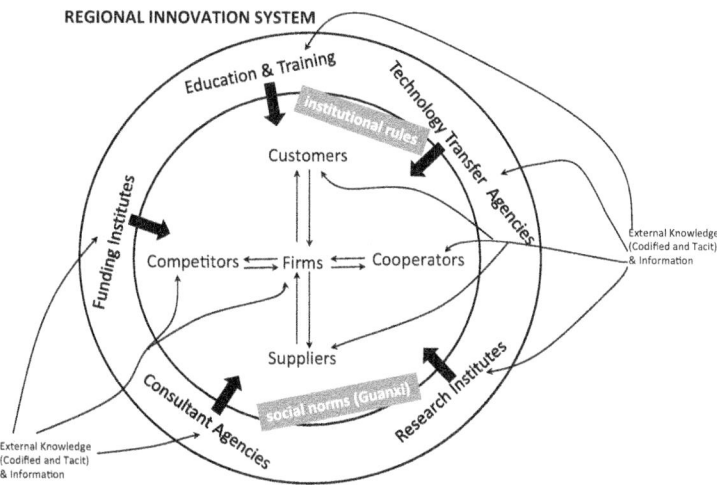

Fig. 1.1 Conceptual regional innovation system

Firstly, the regional specialized clusters in China should consistently source codified and tacit technological knowledge from the distant parent companies and foreign customers, feeding the regional innovation system with new knowledge and information. Therefore, the strategies of foreign affiliates of upgrading the value chain and introducing advanced technology, as well as the local firms' capacity to capitalize on organizational proximity with global lead firms in order to foster innovation, come into the center of the investigation.

Secondly, supported governance infrastructure (which Storper 1995, also refers to as one of the sources for untraded interdependency) should be established to shape the localized cross-fertilization process, tapping into the increasing return on the knowledge spillover sourced externally. Because most of the import technology is concerned with complex products and processes, such as in the electronics and machinery industries, interactive and systemic actions should be in place to ensure fruitful knowledge exploitation. Fromhold-Eisebith (2002) calls it the "regional cycles of learning" that promotes the dissemination of know-how from foreign multinational branch plants. The interactive learning takes place either through vertical linkages (between customers and suppliers) or through horizontal linkages (with cooperators or even competitors). In general, the physical organizations in the governance infrastructure interact with the business sector and support them with necessary information and knowledge. Therefore, the formation of interactive learning activities in the sub-system is crucial for the distribution and joint-exploitation of external knowledge. Moreover, the informal *Guanxi* networks play a role as part of the social rules promoting the interactive learning activities.

In summary, three key terms can be derived from the discussion of the implication of regional innovation system in the Chines context: linkage to external knowledge, interactive learning process and supported governance infrastructure. In Sect. 1.3, the research questions will be formulated to tap into these issues.

1.3 Aims and Research Questions

OECD (2005) points out that the innovation process, rather than innovation results, should become the analytical focus of the innovation studies in developing countries. Based on the previous discussion, innovation studies in China should have a systemic perspective instead of a linear one, which focuses on the distributing and exploiting process of the regional innovation system. Following the previous line of argument, this book aims to contribute to the existing literature on regional innovation systems in three respects:

Firstly, the study aims to explore theoretically and empirically the channels of external knowledge spillovers that are able to trigger the local-scale knowledge spillover. As defined by Cooke (2004, p. 3), a regional innovation system is an open system linked to global, national and other regional systems, and is conductive to knowledge generation, exploitation and commercialization. As stated previously,

1.3 Aims and Research Questions

it is assumed in this study that the formation of a regional innovation system in latecomer regions depends on the regional capacity to disseminate and exploit the external knowledge. Therefore, a starting point in the territorial innovation studies in latecomer regions is an analysis and investigation of the possibility of the triggering effect from the inflowing external knowledge that creates dynamic externalities in the region, on which increasing returns are achieved through interactive learning and systemic innovation.

Secondly, the study aims to expand the understanding of the role of informality in reducing transaction costs further, through to its role in reducing uncertainties and risk faced with innovation activities. Especially in the context of China, the *Guanxi* network, which is widely applied in Chinese business modes, has been proved by many studies to have a positive role in reducing transaction costs (Luo 2002; Zhou et al. 2003; Wu and Choi 2004; Meyer et al. 2009). However, a dichotomous pattern in the application of informal *Guanxi* networks in China might exist. On the one hand, *Guanxi* networks are applied by the local suppliers to sustain reliable supplier-customer relationships as well as to achieve flexible and responsive production. On the other hand, innovation activities are kept within the formal hierarchical framework in the global production network, i.e. the innovation ideas and resources rely heavily on the parent companies or foreign customers. In this study, it is only when the informal *Guanxi* network serves as an important aspect of "regional orders" to incentivize and promote the interactive learning and systemic innovation, that it is considered to contribute to the emergence and performance of a regional innovation system in China.

Finally, this study aims to explore the spatial differences in the pattern of innovation activities. The degree and characteristics of a regional innovation system depend on a specific set of institutions and organizations. Therefore, spatial heterogeneity in the provision of governance infrastructure results in different patterns of innovation activities, which refer to the scope and effect of interactive learning. Moreover, an evolutionary perspective will be applied in this investigation, as the regional innovation system is viewed as an evolving process in which dynamics and inertia consistently emerge with the changing market and technological environment.

In order to achieve the research aim, the following key research questions will be addressed:

Theory-guided questions:

- T1: How and under what circumstances do knowledge spillovers sourced externally trigger knowledge spillovers on the local scale, enabling the formation of regional innovation systems in latecomer export-oriented regions?
- T2: Why do firms undertake interactive learning with external partners in the decision-making and implementing process of innovation activities?
- T3: What is the role of social proximity and organizational proximity in interactive learning activities in latecomer export-oriented regions?
- T4: What leads to the dynamics and inertia of regional innovation systems under different governance infrastructures?

Empirical-guided questions:

- E1: Have local-scale knowledge spillovers come into being to sustain long-term development in the face of a changing and fragile post-crisis global market in the export-oriented Guangdong Province, China?
- E2: Which aspects of absorptive capacity enable the electronics firms to undertake interactive learning with external partners through strategies of using organizational proximity and social proximity in the product innovation process?
- E3: How is interactive learning organized in the burgeoning regional innovation system? To be more specific, does interactive learning embed more in socially proximate networks or in organizationally proximate networks?
- E4: What is the effect of interactive learning in general on innovation outcomes? And what is the effect of interactive learning embedded within socially proximate networks and organizationally proximate networks on innovation outcomes respectively?
- E5: How do regional innovation systems in Shenzhen and Dongguan, China, differ from each other in the scope and effect of interactive learning, considering that the two cities are evolving towards regional innovation systems under different governance infrastructures in the initial industrialization phase?

Policy-guided questions:

P1: What policy implications can be drawn from the previous answers from the theoretical and empirical perspectives in order to enhance the innovation capability of firms and regions in China?

1.4 Outline

The book is organized according to three dimensions: the meso-level investigation, the firm-level investigation and the firm-regional level investigation. Table 1.1 displays the overall layout of the book. Chapter 2 provides a theoretical framework for analyzing the overall impact of knowledge spillovers—within the same industry locally, across different industries locally, and through global linkages—on the performance of innovation and technological upgrading within the context of a latecomer export-oriented region. Based on the stylized facts on technological upgrading in one of the most export-oriented areas, Guangdong Province of China, this chapter further collects empirical evidence of the triggering effect of external knowledge spillover on the local-scale knowledge spillover by applying a meso-scale secondary data set in Guangdong Province.

In order to reveal the pattern of local-scale knowledge spillover, Chap. 3 further explores the micro-firm-level evidence of the upgrading and innovation activities among the electronics firms in the PRD, China. It elucidates the logic behind the interactive process of innovation activities and discusses the role of informal *Guanxi* networks on interactive learning in China. In this chapter, the empirical investigation focuses on whether a wider scope and higher intensity of interactive learning

1.4 Outline

Table 1.1 Schematic overview of chapters

Chapter 1 Introduction: Research context, key concepts and aim	
Empirical investigation	
Meso-level evidence	Chapter 2 Knowledge spillovers and regional innovation: The case of Guangdong Province, China
Firm-level evidence	Chapter 3 Interactive learning and systemic innovation: firm-level evidence from the electronics industry in the Pearl River Delta, China
	Chapter 4 Absorptive capacity, proximity and innovation: Linking up the intra-firm characteristics with Inter-firm linkages
Firm-regional insights	Chapter 5 From globalized production systems to regional innovation systems: Governance and innovation in Shenzhen and Dongguan, China
Chapter 6 Conclusions: Answers, limitations and policy implications	

activities would promote the innovation outcomes. Moreover, initial insight will be provided on the application of informal *Guanxi* networks when electronics firms undertake interactive learning activities in the PRD region.

Chapter 4 is the second study at the firm-level, further strengthening the argument in Chap. 3 on the role of interactive learning for electronics firms in the PRD, China. It extends the understanding of interactive learning within the proximity concept and further investigates the capacity of electronics firms in the PRD to capitalize on social proximity and organizational proximity respectively in the process of product innovation. As technology transfer and learning has relied heavily on organizational proximity to leading global firms ever since the initial industrialization in the PRD, insights into the burgeoning regional innovation system are expected, as firms are gradually taking the initiative to capitalize on social proximity with many other business partners in the process of interactive learning and systemic innovation.

The investigation of the spatial difference with which the electronics firms undertake interactive learning is introduced in Chap. 5. In this chapter, the general regional orders, i.e. the governance infrastructure that incentivizes and supports the systemic innovation at the city-regional level, is the study focus. Moreover, an evolutionary perspective towards governance infrastructure will be taken. Adapted to the Chinese circumstance where the regional innovation system is just burgeoning, the evolutionary lens expands to the transition from governance that supports initial industrialization to the governance that supports the innovation activities. As comparative study is the most important means for fully understanding the function of regional innovation systems and capturing hidden variables that are of interest to its construction (Staber 2001; Doloreux 2002, 2004, Asheim and Coenen 2005), an inter-city comparison of the governance evolutionary paths and the resulting innovation pattern between Shenzhen and Dongguan, China, will be made in order to gain these insights.

On the basis of the previous four theoretical discussions as well as empirical insights, the concluding Chap. 6 will provide answers to the key research questions.

Furthermore, the limitations of this study and future research directions will be reflected upon. Finally, policy implications for further strengthening the innovation capability in China will be discussed.

1.5 Survey Data and Evaluation

Except for Chap. 2, which applies the secondary data in Guangdong Province, China, the empirical data for the rest of the investigation (Chap. 3–5) is a set of standardized questionnaire data on electronics firms in the PRD, China. The electronics industry was chosen because it is not only dominant in the industrial structure in the PRD (Fig. 1.2), but is also facing the greatest upgrading pressure due to rapid technological advance and market change.

The survey targeted electronics firms at three different types of locations for a deeper understanding of different phases of regional development: (1) the first ring city Shenzhen, where the share of the output value is over 47% of the electronics industry in Guangdong province, and where many indigenous firms are thriving; (2) the second ring city Dongguan, where the share of the output value is over 12% and was developing quite rapidly in the late 1990s; (3) the third ring cities represented by Huizhou and Heyuan together, where the share of electronics is smaller, but is now developing due to the expanding, relocating and outsourcing activities in Shenzhen and Dongguan.

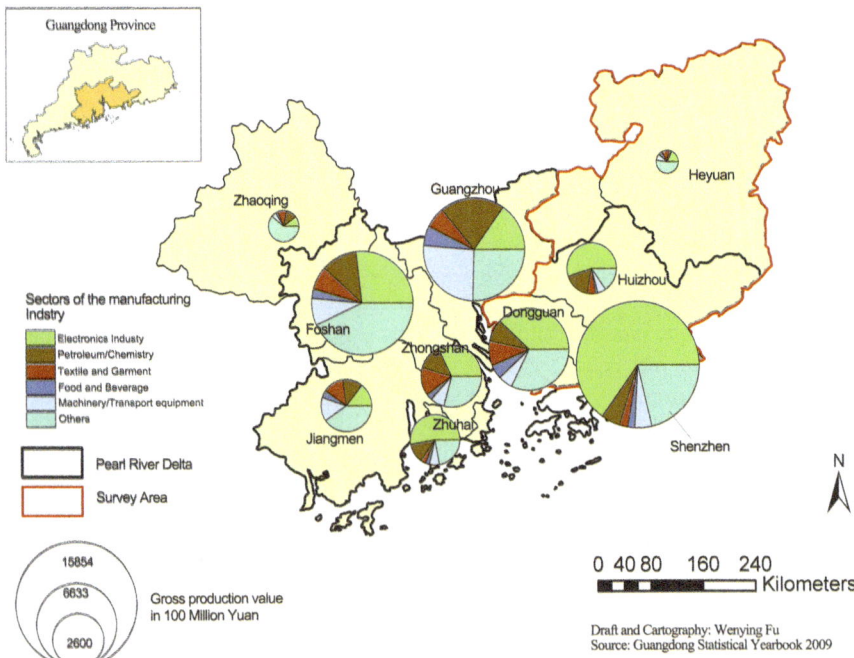

Fig. 1.2 The electronics industry in the PRD, China. (Source: Own draft based on GPBS 2009)

1.5 Survey Data and Evaluation 11

After the discussion with local experts at Sun Yat-sen University and based on the experience in a similar previous survey, the research team decided to conduct the company questionnaires via telephone and mail in order to ensure the feasibility of the survey and validity of the data. It is difficult to get all the questions answered correctly in a limited amount of time in a face-to-face interview, since our questionnaire covers a wide range of company operations from strategic management, marketing, sales, R&D, employment and training (for the final version of the questionnaire please refer to Sect. 7.1 Appendix A), and the respondents needed time to search and consult others while filling out the questionnaires. The telephone and posting method was strengthened by a follow-up process, which aimed to remind and persuade the firms to fill out and send back the questionnaires as well as to fill out unanswered questions and missing items after the questionnaires had been returned.

Our survey was conducted from September to November 2009. In this period, many electronics firms were recovering from the crisis and were quite busy with employing new workers and producing for coming orders. This caused difficulties for our telephone and posting survey because the firms were too busy to pay much attention to us. In order to establish contact with firms, we applied the second method: trade fair visiting. We randomly selected the trade fairs and the firms there, and distributed the questionnaires at the fairs. Because of face-to-face communication, the managers (or people in a high position) felt more embarrassed refusing us than over the telephone. If they were able to answer the questionnaires on site, then we received them back immediately. If they needed some time to consult the boss or related departments about precise information, we asked them to send the answers back to us and carried out the follow-up process.

As for the survey in Shenzhen, the sampling frame was mainly Guangdong Electronics Firm Directory 2010. There are about 2000 Shenzhen electronics firms in this directory, and we applied a random sampling method to select the firms to contact. Within a month, we contacted over 1000 firms, sent 202 questionnaires and received 68 questionnaires back. The research assistants were then asked to recheck the completeness and correctness of the returned questionnaires, and we sent a gift to the responding firms to thank them and also to ensure the success of the follow-up process. Questionnaires were then further improved using the follow-up calls. In order to expand the sample size, we also attempted to contact firms at the fairs. In the spirit of random sampling, we visited three fairs and selected the exhibitors randomly. We firstly visited the 12th International Computer Communication and Consumer Products Expo (Dongguan, China) and received the agreement of 28 firms to fill out the questionnaires (44 Shenzhen firms in total in the fair). We eventually received 23 completed questionnaires. The second one we visited was the 1st China (Shenzhen) International Industrial Fair. There were 145 exhibitors in the electronics field at this fair. We have received the agreement of 32 firms to fill out the questionnaires, and eventually received 29 completed questionnaires. The third fair we visited was the 12th China (Shenzhen) Hi-tech Fair. There were 129 exhibitors in the electronics field at this fair. We received the agreement of 50 firms to fill out the questionnaires, and finished with 47 completed questionnaires. The response rate at the fairs seems to have been better than that of the post due to

Table 1.2 Response rate in different cities and occasions

		Shenzhen	Dongguan	Huizhou	Heyuan
Telephone and posting	Survey firms	68	18	67	11
	Contacted firms[a]	202	31	178	22
	Response rate (%)	34	58	38	50
Fair visiting and posting (or reclaiming on site)	Survey firms'	99	159	–	–
	Contacted firms[a]	110	250	–	–
	Response rate (%)	90	64	–	–

[a] Contacted firms refer to firms which permitted us to send the questionnaires or agreed to fill out the questionnaires at the fairs

face-to-face communication. In total, we received 167 completed questionnaires in Shenzhen.

As for the survey in Dongguan, we firstly contacted the firms which answered our questionnaires in the first phase and established some relationship with us. We contacted 31 firms and had 18 questionnaires completed. We then visited the 12th International Computer Communication and Consumer Products Expo, Dongguan, China. There were about 500 Dongguan electronics firms attending the fair, which is 56 % of all electronics firms in Dongguan. In a sense, attending this fair was a political task for the town governments assigned from the Dongguan city government because 3 C fair is a city card for Dongguan. Dongguan firms received considerable financial incentives from the town governments for attending the fair. Therefore, the representativeness of the Dongguan exhibitors was quite good and ensured the unbiased nature of the fair visiting result. We distributed 250 questionnaires there and received 159 completed questionnaires.

As for the survey in the third ring cities Huizhou and Heyuan, the sampling frame was the Huizhou Electronics Firm Directory 2010 and the Heyuan Electronics Firm Directory 2010. There are 590 and 90 electronics firms in these directories respectively. However, the quality of both directories is fairly low. The repetitiveness is quite high and the accuracy of the information is low. Many telephone numbers do not exist or were constantly engaged. We went through all the available firms. We eventually received permission from 178 Huizhou firms and 22 Heyuan firms to send the questionnaires. 67 questionnaires were returned back from Huizhou and 11 from Heyuan. The overall information on response rate in the research region is summarized in Table 1.2.

The sample distributes quite equally between the first and second rings, while in the third ring, we received a smaller sample. There are two main reasons. Firstly, the total number of electronics firms in Huizhou and Heyuan is much smaller than that of Shenzhen and Dongguan (Table 1.3). The sample size has significantly limited the possible results. Secondly, the firms in Huizhou and Heyuan are much more informal and small-scale, some are even family-run workshops, and it is very difficult

1.5 Survey Data and Evaluation

Table 1.3 Sample distribution (2009). (Data sources: Own survey conducted in DFG SPP 1233 [2009]; GPBS (2010))

	Shenzhen	Dongguan	Huizhou	Heyuan
Survey firms	167	177	67	11
Number of Electronics firms in 2009[a]	1922	860	303	32

[a] It refers to Manufacture of Communication Equipment, Computers and Other Electronic Equipment above designated size include all state-owned firms and firm with over 5 million sales

to establish contact with the bosses or managers. In some cases, the managers could not even understand the questionnaires after our explanation. Besides a lower education level, the cooperating attitude also tends to be lower in the third ring cities.

Due to the financial crisis, the telephone and posting methods we applied in the first phase encountered difficulties. Firms were eager to earn money after a long period of operation pause. The strategic development plan we offered to them afterwards seemed less attractive in this period. The fair visiting and posting (or reclaiming on site) method had a higher response rate under these circumstances. However, the more biased nature of the fair visiting method should be carefully managed. The large share of the exhibitors in the industry in the specific city (such as the Dongguan 3C Expo) can ensure the representativeness of the sample. If this cannot be assured, then the number of fairs visited should be enough to ensure the total quantity of exhibitors to balance the bias.

Table 1.4 presents the comparison between the survey sample and the whole population in Guangdong Province according to firm size and firm ownership. In terms of firm size, the sample and the population do not differ in a significant value of 0.01, but in a significant value of 0.05. However, the difference is quite small (5%) in spite of having a significant value of 0.05. Moreover, it is only possible to stratify the firm size in the sample according to sale and employment, leaving another important criteria, the asset value, unconsidered compared to the official statics in the whole population. If asset value is considered, which is more difficult to achieve, the share of large and medium-sized firms would have been smaller in the sample. In the official statistics, one cannot differentiate between large firms and medium-sized firms, while large firms only take 7% of the whole sample according to the criteria of sale and employment, leaving most of the firms as small and medium-sized. The same goes to the distribution of firm ownership in the sample and the whole population, in which the sample has a slightly larger share (6%) of domestic firms than the whole population in a significant level of 0.05.

Overall, the survey sample in this study is slightly biased towards domestic and medium-sized firms. The biased problem can be partly attributed to the small share of sample in the whole population, which is less than 10%. This statistical problem is explained by the "Jeffrey's paradox" (Jeffreys 1939), in which the population in this case is too large to lead to a significant level even if the difference is uncritical (about 5%). In fact, the χ^2 test is somewhat sensitive in the survey. If it had been managed to reach 18 more small firms in the sample (304 in total), for example,

Table 1.4 Comparison between sample and population based on size and ownership. (Data Sources: Own survey conducted in DFG SPP 1233 [2009]; GPBS (2010))

		Sample ($n=422$)	Population[a] ($N=4645$)
Firm size[b]	Small firms	286 (68%)	3386 (73%)[d]
	Large and medium-sized firms	135 (32%)[c]	1259 (27%)
	[a]$\chi^2=4.765, p=0.029$		
Firm ownership	Domestic firms	217 (52%)	2153 (46%)
	Foreign firms[e]	204 (48%)	2492 (54%)
	[a]$\chi^2=4.181, p=0.041$		

[a] It refers to Manufacture of Communication Equipment, Computers and Other Electronic Equipment above designated size include all state-owned firms and firm with over 5 million sales in Guangdong Province
[b] Large and medium sized firms refers to firms with no less than 30 million Yuan sales, no less than 300 employees and no less than 40 million Yuan assets
[c] In the sample, large and medium sized firms refers to firms with no less than 30 million Yuan sales and no less than 300 employees
[d] Numbers in the parentheses indicate the share of the firms
[e] Foreign firms include wholly foreign-owned, Chinese-foreign equity and Chinese-foreign cooperative firms

then the χ^2 test is not able to sustain in the significant level of 0.1. Similarly, the χ^2 test does not sustain in the significant level of 0.1 if it had been managed to reach 9 more foreign firms in the sample (213 in total).

Therefore, conclusions can be generalized to the whole population in the PRD by focusing the study on the effect and ways of undertaking interactive learning in the innovation process, given the fact that the difference in size and ownership distribution is very small even although in a slightly significant level. Nevertheless, it should be still dealt with cautiousness to compare the size group and ownership group in the sample and to draw conclusions merely by descriptive statistics.

References

Asheim BT, Coenen L (2005) Knowledge bases and regional innovation systems: comparing Nordic clusters. Res Policy 34(8):1173–1190
Asheim BT, Isaksen A (2002) Regional innovation systems: the integration of local 'sticky' and global 'ubiquitous' knowledge. J Technol Transf 27(1):77–86
Aydalot P (ed) (1986) Milieux Innovateurs en Europe. GREMI, Paris
Braczyk HJ, Cooke P, Heidenreich M (1998) Regional innovation systems: the role of governances in a globalized world. Routledge, London
Cooke P (2004) Regional innovation systems: an evolutionary approach. In: Cooke P, Heidenreich M, Braczyk HJ (eds) Regional innovation systems: the role of governance in a globalized world, 2nd edn. Routledge, London, pp 1–18
Cooke P, Gomez Uranga M, Etxebarria G (1997) Regional innovation systems: institutional and organisational dimensions. Res Policy 26(4–5):475–491

References

Doloreux D (2002) What we should know about regional systems of innovation. Technol Soci 24(3):243–263

Doloreux D (2004) Regional innovation systems in Canada: a comparative study. Reg Stud 38(5):479–492

Fromhold-Eisebith M (2002) Regional cycles of learning: foreign multinationals as agents of technological upgrading in less developed countries. Environ Plan A 34(12):2155–2174

GPBS (Guangdong Provincial Bureau of Statistics) (2009, 2010) Guangdong Tongji Nianjian (Guangdong Statistical Yearbook). China Statistics Press, Beijing

Heidenreich M (2004) The dilemmas of regional innnovation systems. In: Cooke P, Heidenreich M, Braczyk HJ (eds) Regional innovation systems: the role of governance in a globalized world, 2nd edn. Routledge, London, pp 363–394

Howells JRL (1999) Regional systems of innovation? Cambridge University Press, Cambridge

Jeffreys H (1939) Theory of probability. Oxford University Press, Oxford

Lundvall BA (1992) National innovation systems: towards a theory of innovation and interactive learning. Pinter, London

Luo Y (2002) Partnering with foreign firms: how do Chinese managers view the governance and importance of contracts? Asia Pac J Manage 19(1):127–151

Meyer S, Schiller D, Revilla Diez J (2009) The Janus-Faced economy: Hong Kong firms as intermediaries between global customers and local producers in the electronics industry. Tijdschr Voor Economische En Soc Geogr 100(2):224–235

Nolan PD, Lenski G (1985) Technoeconomic heritage, patterns of development, and the advantage of backwardness. Soc Forces 64:341–358

Organization for Economic Co-operation and Development (OECD) (2005) Oslo manual: guidelines for collecting and interpreting innovation data, 3rd edn. OECD, Paris

Organization for Economic Co-operation and Development (OECD) (2011) Main science and technology indicators database. OECD, Paris

Porter ME (1990) The competitive advantage of nations. Free Press, New York

Revilla Diez J (2000) The importance of public research institutes in innovative networks-empirical results from the Metropolitan innovation systems Barcelona, Stockholm and Vienna. Eur Plan Stud 8(4):451–463

Smith K (2000) Innovation as a systemic phenomenon: rethinking the role of policy. Enterp Innov Manage Stud 1(1):73–102

Staber U (2001) The structure of networks in industrial districts. Int J Urban Reg Res 25(3):537–552

SCPRC (State Council of the People's Republic of China) (2006) Guojia Zhongchangqi Keji Fazhan Guihua (2006–2020) (National medium and long term Science and Technology Development Plan (2006–2020)). Beijing, China

Storper M (1995) The resurgence of regional economies, ten years later: the region as a nexus of untraded interdependencies. Euro Urban Reg Stud 2(3):191–221

Wu WP, Choi WL (2004) Transaction cost, social capital and firms' synergy creation in Chinese business networks: an integrative approach. Asia Pac J Manage 21(3):325–343

Zhou X, Li Q, Zhao W et al (2003) Embeddedness and contractual relationships in China's transition economy. Am Sociol Rev 68(1):75–102

Chapter 2
Knowledge Spillovers and Regional Innovation: The Case of Guangdong Province, China

Abstract This chapter aims at analyzing the impact of knowledge spillovers through external channels, namely FDI and trade on the regional innovation and upgrading in Guangdong province, China by using the panel data of 21 municipalities for the period of 2001–2012. The results show strong evidence of external knowledge spillover as effective trigger of local-scale knowledge spillover in the latecomer regions. The external knowledge spillover is primarily working through the mechanism of FDIs rather than import at the investigation period. Furthermore, it demonstrates that the impact of external knowledge spillover is moderated by the degree of industrial diversity at the city scale, which means that more diversified areas benefit from FDI and import to a greater extent. Overall, the empirical investigation suggests that external knowledge spillover is not an automatic process, but is rather closely related to the investment stock, the degree of embeddedness and the absorptive abilities of local firms in the long run. At the end, this chapter points out the future study should go further to explore microeconomic aspects of innovation at the firm-level.

2.1 Introduction

In the field of economic geography and regional development, growing research attention has been paid to the role of knowledge spillovers in generating endogenous growth and determining economic development. Knowledge spillover is a typical urban phenomenon. Going beyond explaining the mere existence of cities, as what static externalities theory does, knowledge spillovers explain the growth of cities (Glaeser 1999). Aside from explaining regional economic growth from the perspective of cost savings on transportation and intermediate inputs (Hoover 1937;

The chapter is the second version of a published paper *Knowledge Spillovers and Technological Upgrading: The case of Guangdong Province, China, Fu W, Revilla Diez J, Asian Journal of Technology Innovation 18/2*, Copyright © 2011, Taylor&Francis. By extending the previous investigation period (2000–2008) to further cover the post-crisis period (2001–2012), the underlying dynamics for knowledge spillover being concluded in this chapter has changed in certain aspects compared to the first version.

Carlton 1983; Krugman 1991), knowledge spillovers theory reinterprets externalities in a dynamic way and argues that innovation investment bears increasing returns because it contributes to a general stock of knowledge upon which neighboring firms or latecomer firms can develop (Jacobs 1969; Romer 1986; Lucas 1988; Glaeser 1999).

Two kinds of externalities, namely intra-industry and inter-industry externalities, on the local scale are vital to the growth of cities (Glaeser et al. 1992). However, research findings on knowledge spillovers are heretofore mixed (Feldman 2000). On one hand, it has been proven that specialization stimulates growth, in which knowledge spillover in the same industry is active (Henderson and Cockburn 1996; Henderson 2003). On the other hand, industrial concentration is suggested to hinder growth in some way (Miraky 1994), and the positive impact of diversity has been otherwise proven. Glaeser et al. (1992) discover that knowledge spillovers across industries—rather than within the same industries—boost employment in a period of deindustrialization, particularly in traditional industrial cities. Feldman and Audretsch (1999), and Rosenthal and Strange (2003) also confirm similar benefits of diversity.

The ambiguous pattern of research is somewhat related to different studied objects positioned in different context of time and space (Combes 2000). Comparing the mixed cases, a hidden rule has emerged: knowledge spillovers within the same industry primarily induce incremental innovation, whereas knowledge spillovers across industries are conducive to disruptive innovation. Neffke et al. (2011) latest discourse on the impact of different kinds of knowledge spillovers on the industrial life cycle develops this statement. It is suggested that knowledge spillovers take place across industries when industries are young and rejuvenating, whereas knowledge spillovers within the same industry is more prevalent when industries grow and mature. However, the above literature lacks an open perspective in the era of globalization. Branstetter (2006) finds that Japanese FDIs in the United States result in two-way knowledge spillovers between the two countries. Boschma and Iammarino (2009) conduct a systematic measurement of knowledge spillovers on the local and global scales in Italy, and find that a high variety of traded goods into the region contributes to regional economic growth. Furthermore, knowledge spillovers between developed and developing countries are intensely examined and considered as a key mechanism for conditional convergence in the global economy. The positive relationship between trade and growth is confirmed in studies asserting the knowledge spillovers from the industrial North to the developing South (Coe et al. 1997; Falvey et al. 2004). Javorcik (2004) proves productivity spillovers induced by FDIs across industries in Lithuania, one of the transition economies in Eastern Europe, through forward and backward linkages with foreign affiliated firms. China's fast growing rate during the recent decades has been also suggested by some scholars to be partly attributed to the active integration into the global production networks (Lemoine and Unal-Kesenci 2004; Yeung 2009), although the effect of technology spillover is debated (Wei et al. 2009; Wang and Lin 2013).

Based on the literature review, this chapter argues that with access to external advanced knowledge and requisite absorptive ability on the local scale, it is very

2.1 Introduction

possible for latecomer regions to seize the inflow of external knowledge, thereby triggering knowledge spillovers on the local scale, which are likely to lead to the formation of regional innovation systems to sustain long-term economic growth. In addition, the missing link between external knowledge spillover and localized knowledge spillover in the literature is addressed in the chapter, with theoretical and empirical investigation on the moderating role of urban industrial structure on the effect of FDI and import on innovation. Overall, the chapter contributes to the literature in two respects. First, I put knowledge spillovers on both local and global scales in the latecomer context within an integrated theoretical framework, and discuss how knowledge spillovers on the global scale trigger knowledge spillovers on the local scale, and also interacts with local structure. Second, evidence on knowledge spillovers, which underlies innovation and economic growth in the modern economy, is further collected within the latecomer context. After the global financial crisis and the gradual recovery of East Asian economies[1], it is important to examine whether the local dynamics of economic development, such as active knowledge spillovers, has come into shape to sustain long-term development in the face of a changing and fragile post-crisis global market.

The study area is Guangdong province in South China. The province is selected based on two reasons. First, Guangdong has developed quickly after the opening of the Chinese economy by having successfully attracted labor-intensive production. Latest statistics in 2012 show that FDI in Guangdong accounts for 15 % of the national total, ranking the second among the regional peers, and Guangdong's total import and export volume accounts for about 29 % of the national volume (GPBS 2013). Second, since China's transition from a planned economy to a market economy in 1978, technological activities, such as investments in equipment renewal, process innovation, and product upgrading, have become increasingly prevalent among enterprises in Guangdong (Wang 2008). These factors justify the choice of Guangdong for testing the existence and impact of knowledge spillovers after decades of development.

The chapter has the following structure. Section 2.2 collects stylized facts on technological upgrading and innovation in Guangdong province, China. Section 2.3 provides a theoretical framework for analyzing the overall impact of knowledge spillovers—within the same industry locally, across different industries locally, and through global linkages—on the performance of innovation within the context of a latecomer region, and further elucidates how urban industry structure influences the impact of external knowledge spillover on regional innovation. The section also derives hypotheses for empirical testing. Section 2.4 describes the variable design and the model specification. Section 2.5 reports the results. Finally, Sect. 2.6 provides the conclusion and discusses ways to further extend our understanding of knowledge spillovers.

---

[1] See Willem Thorbecke, guest edition "East Asian production networks, global imbalances, and exchange rate coordination". Econbrowser: analysis of current conditions and policy, 19 October 2009. http://econbrowser.com/archives/2009/10/east_asia_the_g. Accessed 13 May 2014.

2.2 Technological Upgrading and Innovation in Guangdong Province, China: Some Stylized Facts

Economic activities in Guangdong province, in particular the core region—the Pearl River Delta (PRD), are characterized as processing and compensation trades (*sanlaiyibu*). In spite of the relatively low added value in the global value chain, content of processing trade in Guangdong province has in actuality upgraded from low-tech products, such as garment and shoes, to high-tech products, such as electronics. The share of high-tech products, mainly the electronics products, in total exports in Guangdong has increased from 19 to 39% from 2000 to 2012, whereas the share of leather, textile and footwear products decreased from 23 to 11%. Upgrading of processing trade in industry categories clearly indicates the greater ability of PRD firms to understand, absorb, and process more complex products. Moreover, the share of processing and compensation trade has gradually fallen from 78 to 56% during the period 2000–2012, implying more independency of industrial production.

Because the organization of electronics product is quite fragmented, suppliers in the emerging countries are distributed in different tiers of the global value chain in terms of skills and technological level. Therefore, deeper investigation into structural changes in primary high-tech trade products is needed. The processing trade involves large volume of components import (particularly the key high-tech component) and final goods export. As shown by the structured trade data in Tabel 2.1, data processing equipment and mobile phones did not appear among the main high-tech export product catalog in 1997. After 2001, export trade of these two electronics commodities has surged, account for 50% and 34% of the major exports in 2012, respectively. The rising export of the data processing equipment and mobile phones should complement with, given China's relatively lower technological capabilities, the importation of more semi-finished components, especially the integrated circuit (from 55 to 80%). On the other hand, the increasing import of other components, primarily part of semiconductor devices and circuit protection fixtures, which are complex electronic components used in integrated circuits, demonstrates the firms' ability to design and produce their own integrated circuit. The processing of these components clearly indicates a deeper understanding of the principle of circuit functions and requires an increasing ability in the field of circuit design adjustments for different purposes. The inference that firms are strengthening their capability to adapt and improve imported technology is further corroborated by the accelerating international competitiveness of the IC products, with its share rising from null to 13% of the total exports in Guangdong province. In addition, the import of kinescopes—the core technological component of TV sets—has decreased to a great extent. Combining this with the increased export of color TV, it can be concluded that firms in Guangdong are already capable of producing TV sets on their own; this signifies to a certain extent technological upgrading.

New product development is another important aspect of technological upgrading. Under China's statistical standards, new products refer to brand new products

2.2 Technological Upgrading and Innovation in Guangdong Province, China

Table 2.1 Structural change of electronics industry trade in Guangdong. (Data Sources: GPBS (1998, 2008 and 2013))

Final and intermediate goods	Value of trade (unit: $ 10,000 US)		
	1997	2007	2012
Imports			
Final electronic goods	49,493	993,009	1,479,630
Data processing equipment	44,547 (90%)	989,411 (99%)	1,479,079 (99%)
Exchange of telephone and telegram	322 (1%)	–	–
TV sets	4,624 (9%)	3,598 (1%)	551 (1%)
Electronic components	443,483	6,283,113	9,258,958
Integrated circuit (IC) and parts of electronic components	245,155 (55%)	4,920,595 (78%)	7,404,316 (80%)
Part of semi-conductor devices	76,960 (17%)	640,546 (10%)	913,381 (10%)
Protection fixtures of circuit	57,416 (13%)	545,889 (9%)	735,005 (8%)
Kinescope	20,017 (5%)	13,669 (1%)	1,660 (0.1%)
Electric wire & cable	43,935 (10%)	162,414 (2%)	204,595 (2%)
Exports			
Final electronic goods	297,181	6,248,260	10,243,144
Data processing equipment	–	4,110,387 (66%)	5,176,401 (50%)
Mobile phone	–	923,567 (15%)	3,476,778 (34%)
Electric calculator	32,527 (11%)	64,765 (1%)	39,955 (1%)
Telephone set	81,574 (27%)	287,209 (4%)	161,260 (2%)
Loud speaker	36,515 (12%)	192,603 (3%)	330,183 (3%)
Hi-fi stereo component system	17,510 (6%)	381,326 (6%)	290,167 (3%)
Radio set	28,394 (10%)	–	–
Color TV sets	29,125 (10%)	284,235 (4%)	490,450 (5%)
Camera	71,536 (24%)	4,168 (1%)	277,951 (2%)
Electronic components	75,831	882,782	2,608,407
Integrated circuit (IC) and parts of electronic components	–	259,199 (29%)	1,613,685 (62%)
Static current transformer	75,831 (100%)	623,583 (71%)	994,722 (38%)

Number in brackets indicates the share (%) of this product in sub-category export or import products

that utilize new technology or new design ideas, or greatly improved products in terms of structure, material, or processing methods, all of which would either significantly enhance the performance of the products or expand their functions. Based upon the North-South spillover assumption (Coe et al. 1997), firms in latecomer countries have a large room for learning as well as adapting technology developed by industrialized countries to improve performance. Hence, within the latecomer

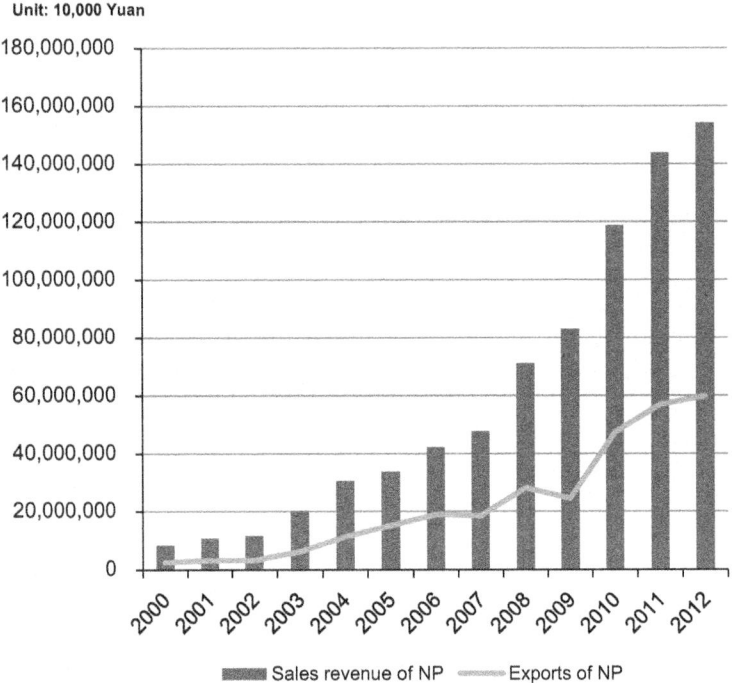

Fig. 2.1 New product development in Guangdong. (Data source: GPBS (2001–2013, annual))

context, new product development is a more market-oriented indicator than patents that represent the technological frontier. Figure 2.1 shows that the sales revenue of new products in Guangdong increased more than 18 times over the decade during 2000–2012, among which the export of new products increased 23 times from 2000 to 2012.

New product development activities are not evenly distributed in Guangdong province; some cities score high in product innovativeness, while some others are lagging behind. The most innovative city, Shenzhen, has achieved about 600 times as many new products as the least one—the mountain city Yunfu. Internal learning efforts of firms constitutes a necessary requisite for achievements in new product development; here, correlation analysis of Spearman's rho is used to calculate the relationship between new product sales and export, and R&D expenditures from 2001 to 2012 across 21 municipalities in Guangdong. It is shown that the new product sales revenue is significantly and positively correlated (0.922) with firm R&D expenditure. Likewise, significant and positive correlation (0.865) is also found between new product export and R&D expenditure (Tabel 2.2).

R&D activities involve both efforts in pushing technological frontiers and improving existing knowledge. International comparative studies indicate that over 1% of R&D expenditure in GDP signifies that a country/region has passed the phase of basic technological introduction and application, and has developed the

2.2 Technological Upgrading and Innovation in Guangdong Province, China

Table 2.2 Spearman's correlation of new product performance with R&D expenditure. (Data sources: Calculation based on data compiled from GPBS (2002–2013, annual))

Spearman's rho correlation		Firm R&D expenditure
New product sales	Correlation coefficient	.922[a]
	Sig. (2-tailed)	.000
New product export	Correlation coefficient	.865[a]
	Sig. (2-tailed)	.000

[a] Correlation is significant at the 0.01 level (2-tailed). List wise $N=147$

ability to absorb and assimilate technology. In 2000, Guangdong's R&D share in GDP sharply increased from just 0.2 % in 1995 to 1.1 %, 2 years earlier than the national average reaching above 1 %. In 2012, R&D expenditure in Guangdong has reached 124 billion Yuan, taking up 2.2 % of the GDP. Compared to the industrialized countries (e.g. the United States, 2.7 %; Japan, 2.7 %; OECD countries, 2.3 %; newly industrialized countries or regions like South Korea, 3.5 %; Taiwan, 2.6 %; and Singapore, 2.3; See OECD 2008), Guangdong is catching up and approaching the high standard. Speaking of the quality, however, R&D activities in Guangdong mainly involve learning efforts in assimilating and improving technologies from industrialized countries.

In order to reveal the character of R&D activities in Guangdong, structural comparison with other developed provinces in China and the national average is made (Tabel 2.3). First, R&D intensity in Guangdong outperforms the national average, but is still much lower than Beijing and Shanghai. This may be attributed to the insufficiency of technological investment in universities and research instituts, as well as the absence of global lead firms devoted to R&D activities, compared to Beijing and Shanghai (Kroll and Tagscherer 2009). Second, technological upgrading in Guangdong is largely market-driven; it is mainly led by private investment other than public one. The percentage of firm investment in R&D in Guangdong is 87 %, which is higher than the national average (74 %) and all other developed regions. Third, technological changes in Guangdong are mainly driven by test and development activities, implying that Guangdong's technological activities are incremental rather than radical.

In the 2000s, including the tough time in the aftermath of the worldwide financial crisis, competition for low-cost regions had been intensifying compared to the 1990s, but still Guangdong showed its learning ability to renew and upgrade products gradually, leading to the sustainable competitiveness of its export sector. Technological upgrading activities in Guangdong are characterized by firm-led assimilation and improvement undertakings. Nevertheless, the mechanism favoring rapid technological changes is missing due to the absence of research excellence of universities and public institutions. Aside from the internal efforts of firms, knowledge spillovers that transfer know-how and induce learning by doing in firms are supposed to be important for the dynamic self-sustaining technological progress of Guangdong. The following sections would further explore the nature of knowledge spillovers both theoretically and empirically.

Table 2.3 National comparison of technological indicators (2012). (Data sources: Calculation based on data compiled from CSSB (2013), GPBS (2013), SMBS (2013), BMBS (2013), JPBS (2013) and ZPBS (2013))

	National statistics	Guangdong	Shanghai	Beijing	Jiangsu	Zhejiang
R&D expenditure (unit: billion Yuan)	1029.8	123.62	67.93	106.33	128.80	72.26
R&D expense (percentage in GDP) (%)	2.0	2.2	3.4	5.9	2.3	2.1
# Firm investment in R&D (%)	74	87	63	40	62	81
# Investment in basic research (%)	5	2.6	7.2	11.8	2.5^1	2.4
# Investment in application research (%)	11.2	7.2	13.5	22.8	6.5^1	5.4
# Investment in test & development (%)	83.8	90.2	79.3	65.4	91^1	92.2

2.3 Impact of Knowledge Spillover on Innovation: An Integrative Perspective from the Local and the Global Scale

In the management literature, innovation is mainly determined at the firm level. Firms that set long-term technological development strategies and devote much of their resources to R&D activities are assumed to achieve better innovation outcomes. The success of the Korean industry follows the logic of internal capability for technological upgrading and innovativeness. The Korean government has created favorable policies in "preferred industries", which has ensured efficient scale economies through mergers, project-specific financial support, and domestic market protection (Chang 1993). Combined with their own R&D efforts, Korean companies have ultimately upgraded to a higher position in the global value chain and established a modern industrial system led by the automobile and electronics industries.

However, firm-level internal efforts cannot explain two phenomena. First, regions with many small firms, such as the Third Italy, which lack the financial ability and economies of scale to support intense R&D activities, perform fairly well in innovation, particularly in incremental innovation (Storper 1995). Rather, mutual trust among firms constitutes the fundamental basis of long-term cooperation, facilitating knowledge exchange and stimulating growth. Although the ideological concept for "small enterprise spatial system" has been under constant suspicion, especially in the post-2000 period, its success as a geo-historical formation has presented itself as a perfect case for the external sources of firm competitiveness (Bianchi 1998; Boschma and Lambooy 2002). Second, firms with the same endowment and efforts in innovation usually perform differently in different locations. All of these suggest that the external environment plays an important role in determining the innovation performance of firms.

To deal with it, I focus on three perspectives on knowledge spillovers. These perspectives are concerned with innovation externalities achieved through knowledge spillovers that enable firms to benefit from each other's internal efforts. The first two knowledge spillovers, which take place within and between industries on the local scale, have been properly modeled and surveyed by many scholars (Loury 1979; Glaeser et al. 1992; Asheim 2000; Neffke et al. 2011). The third knowledge spillover deals with externalities on the global scale (Grossman and Helpman 1990; Branstetter 2001; Javorcik 2004; Branstetter 2006; Parrado and De Cian 2014), which are highly important for firms in latecomer countries, where spillovers from neighboring firms are quite limited.

The flow of ideas is intrinsic to the new knowledge production system that underpins economic growth (Lucas 1988). Glaeser et al. (1992) suggest that people agglomerate in high-rent cities because they benefit from learning opportunities. In this respect, it is assumed that physical proximity facilitates information transmission. Marshall-Arrow-Romer's (MAR) externalities and Jacobs' externalities focus on spillovers on the local scale.

The MAR externalities were developed by Arrow (1962) and Romer (1986) based on Marshall's (1920) agglomeration theory. Marshall's agglomeration theory suggests that firms in the same industry agglomerate to benefit from knowledge spillovers. Moreover, their agglomeration is a cost-saving strategy in their search for intermediate goods and skilled workers. Arrow further expands the theory by stressing the role of knowledge spillovers between workers within the same working area, and argues that experience and learning by doing are vital to endogenous technical changes. Romer's work asserts that knowledge stock generates increasing returns. Thus, specialization is conducive to long-run growth. According to Glaeser's (1999) argument on learning in cities, cities filled with young people who primarily learn from skilled members in their own industries tend to be specialized. The concept of proximity further explains the function of specialization. Geographical and cognitive proximity works in the knowledge spillover process. Cognitive proximity in the same industry assures the basic absorptive ability of firms to assimilate and improve transmitted knowledge, and the exchange of knowledge is facilitated by geographical proximity. Generally, knowledge spillovers within industries accelerate the generation of know-how and lead to incremental innovation, which underlines the success stories of many traditional industrial districts (Amin 2000). The success of the computer chip industry in Silicon Valley corroborates the positive relationship between specialization and knowledge spillover (Saxenian 1994). Skilled workers meet, chat, and eavesdrop, and labor flows across firms, thereby spreading ideas and know-how quickly among co-locating firms.

Meanwhile, Jacobs (1969) holds a different opinion on the way knowledge spillovers take place. Jacobs' externalities stress the diversity of industries as an important factor inducing human capital spillovers and the formation of new ideas. Unlike Arrow's statement that human capital is enhanced by interaction in the same line of work, Jacobs suggests that cross-fertilization across different lines of work enhances human capital in cities. The vivid examples given by Jacobs are new forms of adhesive tapes developed by a sand mining company, brassiere invented by a dress maker in New York, and Japanese bicycle repair shops gradually moving into bicycle manufacturing. Boschma (2004) further develops this argument with evolutionary thinking, stressing that diverse but related knowledge stock is a key factor that determines the effective interaction of actors in locations and prevents the negative "lock-in" effect of specialization. In other words, diversity brings two benefits: knowledge spillovers across different industries, and the portfolio effect that makes regions resilient to external shocks. However, a diversified economy may lead to the lack of focus on general services, such as administrative services, advertising, and legal consultation (Neffke et al. 2011). On the other hand, a specialized economy enables local governments and professional service providers, such as marketing and accountancy firms, to create tailor-made services.

Neffke et al. (2011) discuss the relationship between the industrial cycle and externalities. They conclude that MAR externalities are vital to growing and maturing industries where technological activities focus on improvement and adaptation, whereas Jacobs' externalities are pivotal to emerging industries where technological activities focus on innovation and change. Under MAR externalities, experience and learning by doing play a vital role only when specific technological standards and paradigms are established in the industry. Glaeser (1999) also argues that diversification tends to be lower in cities when imitation is more feasible. In contrast, at the onset of new industries, various new products emerge in the market to compete fiercely because standardization does not yet occur (Gort and Klepper 1982). Therefore, for an infant industry experiencing rapid technological changes, the need to absorb different fields of knowledge to spur ideas and innovations is imperative, and Jacobs' externalities are more important in this case.

Based on the foregoing discussion on knowledge spillovers on the local scale, the first hypothesis is drawn:

Hypothesis 1 In many latecomer regions where improvements and adaptations are prioritized, knowledge spillovers within industries, which stimulate the process of learning by doing, contribute more to innovation than knowledge spillovers across industries.

At the early phase of industrialization, knowledge spillovers on the local scale can be hardly realized due to a weak local industrial base and an unbalanced knowledge distribution among firms. Summarized from the literature, knowledge spillovers are realized mainly through four mechanisms: inter-firm collaboration, inter-firm cooperation, spin-off, and talent mobility (Boschma and Lambooy 2002; Power and Lundmark 2004; Parthasarathy and Aoyama 2006). In the first two mechanisms, firms should have developed their own core technological capabilities, enabling collaboration with customer or supplier, as well as cooperation with firms producing similar products. This ensures the reciprocal exchange of respective knowledge stocks. If firm-level technological capabilities are not fully and consciously developed, and those between firms are not equivalent and supplementary, firms would be less inclined to exchange knowledge due to the lack of mutual benefits.

2.3 Impact of Knowledge Spillover on Innovation

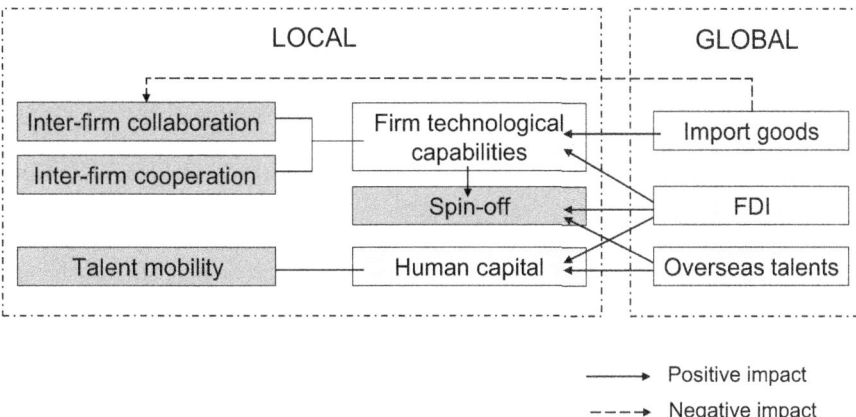

Fig. 2.2 How external knowledge spillovers trigger local knowledge spillovers

Similarly, spin-off activities happen only when parent firms have mature technological paradigms and particular sets of technological capabilities, enabling key employees to exploit the existing knowledge by establishing new organizations. For the last mechanism, i.e. transfer by skilled workers and talents, its effective functioning is also determined by a high level of educational and professional skills, which is difficult to achieve at the early phase of industrialization due to the high share of employment in the agricultural sector.

External knowledge spillovers can trigger knowledge spillovers on the local scale through the abovementioned mechanisms. The importance of external linkages in regional innovation has attracted due attention within the latecomer context (Coe et al. 1997; Humphrey and Schmitz 2004; Falvey et al. 2004; Javorcik 2004; Revilla Diez and Kiese 2006; Oro and Prichard 2011). In latecomer countries, external knowledge spillovers take place in three ways: trade, FDIs, and overseas talents. Figure 2.2 illustrates the sequence of how external knowledge spillovers trigger local-scale knowledge spillovers, in which the four knowledge spillover mechanisms are marked with darker grey. The mechanism of local knowledge spillovers is either directly or indirectly triggered by external knowledge.

First, trade transmits knowledge embedded in imported goods (Parrado and De Cian 2014). Given the high share of components import in Guangdong, competition in the final goods market is not likely to emerge. Imports of components, on the other hand, is able to increase the variety of inputs, finally leading to enhanced productivity of the final product sector (Romer 1986). What's more important, advanced imported components serve as an important source of technology acquisition (Lemoine and Unal-Kesenci 2004), boosting intra-firm learning as a process of shaping the technological capabilities of firms. The enhancement of technological capabilities further enables inter-firm collaboration and cooperation, inducing knowledge spillovers between firms, and further triggers spin-off activities that exploit the value of accumulating knowledge stock. However, firms that rely on imported components are also less inclined to establish local vertical linkages, reducing the possibility of inter-firm collaboration and the induced knowledge spillover. Therefore, the impact of component import on regional innovation bears a dual character; it can either facilitate knowledge spillover by enhancing firm capability and the knowledge stock in the localities as a whole, or inhibit knowledge spillover by weakening the inter-firm linkages.

Second, FDI generates a globally integrated production network in latecomer countries and assists related firms within the network, such as subsidiaries, original equipment manufacturer (OEM) suppliers, and non-OEM suppliers, by introducing, interpreting, and instructing production know-how and product-specific technologies. As such, technological capabilities are strengthened, enabling the functioning of the three mechanisms of local knowledge spillovers as shown in Fig. 2.2. Ivarsson and Alvstam (2005) suggest that geographical proximity to Volvo enables local auto-suppliers to absorb external technology successfully. In addition, FDI brings about the effect of training. Through training, the skill level of the local labor market is enhanced. Although the local firms are in a disadvantaged position in the war for talent vis-à-vis the foreign firms, it is still possible that the employees, who once worked in the foreign subsidiaries, reach the career ceiling and seeks more development room in other local start-ups, and carry knowledge to these firms. Employees with improved market and managerial knowledge can even seize one of the technological opportunities in their parent companies and establish new firms to exploit the market potential of a particular technology, triggering the functioning of the spin-off mechanism.

Last but not the least, the inflow of overseas talents not only increases the quality of human capital on the local scale but also induces entrepreneurial activities under an effective incentive framework offered by the local government. Incentives are provided to attract overseas entrepreneurs, leading to the introduction of latest technologies to the local market. Overseas talents may even establish dynamism between developing home countries and the developed world thorough their personal networks in both places. This is made possible by the formation of "technological communities," a concept introduced by Saxenian and Hsu (2001) in the case of Hsinchu high-tech cluster, and further elaborated by Parthasarathy and Aoyama (2006) in the case of Bangalore software cluster.

However, the trigger effect of external knowledge spillovers does not take place automatically. Interaction of global- and local-level knowledge spillovers underpins the success of upgrading and innovation. The impact of FDIs and imported good on innovation depends on the investment stock, degree of embeddedness, and absorptive ability of local firms.

At the early phase of FDIs, global lead firms seek optimal locations where they can conduct parts of their complex production processes—to take advantage of cheaper production factors and utilize location-specific resources. Their previous investment aimed at cost reduction. At this stage, global firms take a "stand-alone" attitude and exert minimal influence on local capabilities upgrading. Foreign investment at later phase tends to aim at utilizing local resources and exploiting location-specific know-how; thus, investments expand from simple production to basic R&D activities (Maskell and Malmberg 2007). Today, many R&D activities

2.3 Impact of Knowledge Spillover on Innovation

are becoming standardized and codified, facilitating the geographical dispersion of R&D units. These R&D units, given their geographical proximity, support large-scale production activities in latecomer regions. At this phase, the impact of FDI on technological upgrading and innovativeness of local industries is becoming increasingly larger as the scale of investment and embeddedness within the local production networks deepens.

The requirement of which imports trigger knowledge spillovers to latecomer countries is more demanding than what FDIs require because the learning process through imports needs proactive actions. The learning processes from imported goods require the firms to identify strategically, assimilate, and adapt. In the short run, however, imports of technologically advanced components enable the firms to upgrade the products, boosting the regional innovativeness in the form of new-to-the-firm incremental innovation. In the long run, absorptive capacity comes into play, determining the assimilation of imported goods and the effective transformation of knowledge learned from imported goods into the technological capabilities of firms (Cohen and Levinthal 1990). Hence, local firms can only benefit from imported goods when their own absorptive abilities are developed to a certain level. Otherwise, the waning inter-firm connection due to overreliance on imported components would take over the effect on enhancing firm capability, leading to the negative effect of import trade in the region.

Following the above argument, the second hypothesis is drawn:

Hypothesis 2 External knowledge spillovers function through the ways of imported goods, FDIs, and overseas talents; however, realization of external knowledge spillovers on innovation does not occur automatically. The long-term impact of external knowledge spillover though FDI and import on regional innovation depends on investment stock, degree of embeddedness, and the absorptive capacity of local firms.

The external perspective on the global pipelines for knowledge sourcing should not overlook the coordination of local-level resource and structure (Humphrey and Schmitz 2004). In this chapter, I focus on the urban industrial structure, i.e. the specialized local economy in MAR externalities and the diversified local economy in Jacobs' externalities, and how they influence the impact of FDI and import on regional innovation.

Under weak intellectual property protection, foreign firms are less inclined to engage in knowledge exchange and technology transfer activities vis-à-vis the local firms. Unlike the local firms, which have formed flexible inter-firm networks through personal relationships as a substitute for efficient formal institutions (Peng et al. 2008), the absence of social assets in the local clusters prohibit the foreign invested firms from exchange and dissemination with a need to reduce excessive imitation from other firms. This situation is more severe in a specialized local economy, in which incremental innovation is dominant and the cost of imitation is quite low. Also for a highly specialized lobar pool, FDI endangers the upgrading prospects of domestic firms by attracting the key technological staff, thus creating rapid staff turnover and discontinuous process of capability accumulation in the domestic sector (Parthasarathy and Aoyama 2006). In a more diversified urban economy, on the other hand, foreign firms are more willing to exchange with firms in the related industry to tap into the local knowledge stock, without much fear for fierce competition with similarly imitated products.

As elucidated in the second hypothesis, the impact of imports is greatly determined by the absorptive ability of firms, and the literature on firm clustering asserts that leading firms act as gatekeepers that determine the effective diffusion of import knowledge to the clustered firms (Owen-Smith and Powell 2004; Giuliani 2005). In this sense, the absorptive ability of leading firms determines the impact of imports on knowledge spillovers on the local scale. Due to diseconomies of scale in more diversified economies, the technological absorptive capacity of leading firms in a diversified location is weaker than those in a specialized one, rendering them in an inferior position to adeptly apply imported components to upgrade the product lines. In contrast to the short-run negative effect of diversified economy, it is beneficiary in the long run by reducing the local firms' overreliance on imported technology. In addition, the firms would find more possibilities in a diversified city to exploit market opportunities in related industries by making new combinations from the advanced imported components after a certain period.

Finally, the third hypothesis is formulated concerning the moderating role of urban industrial structure for the causal relationship between external knowledge spillover and innovation.

Hypothesis 3 The relationship between external knowledge spillover and innovation is moderated by urban industrial structure. In the long run, innovation performance increases with external knowledge spillover but at a faster rate in a diversified city.

2.4 Model Specification and Data

The empirical testing of knowledge spillovers is mainly conducted using three methods. The first method includes the effects of knowledge spillovers as part of the technology term into a regional production function by using city-industry data (Glaeser et al. 1992; Feldman and Audretsch 1999; Neffke et al. 2011). The second method constructs a knowledge function (patents or new products) to catch the "technological leakage" of private (firms within or across industries) and public research institutes (Jaffe 1989; Feldman 1994). The third method traces the time and spatial scale of patent citations (Trajtenberg 1990; Jaffe et al. 1993).

According to the pilot round of 1992 Community Innovation Survey in the Netherlands, there are basically three categories of innovative indicators: R&D, patent application and new product efforts. It has been suggested that new product introduction is an indicator that is more direct and market-oriented than patented inventions (Griliches 1990; Feldman and Audretsch 1999; Kleinknecht et al. 2002). In the latecomer context, knowledge assimilation and adaptation is more prevalent than knowledge production, which makes the use of patent data in measuring innovativeness quite irrelevant. Although Cheung and Lin (2004) find a positive spillover effect by FDIs on domestic patent application in major coastal provinces of China, this spillover effect is strongest in the field of minor innovations, in particular external design innovations. According to an innovation survey on 8962 firms in Guangdong in 2008 (See Wang 2008), product innovation is as important as process innovation. Therefore, I use the share of new product in total industrial output value as the proxy for innovativeness in the research region.

With regard to the specific function form, the first method of testing knowledge spillovers mentioned above cannot be utilized because city-industry data on external linkages, such as those on FDIs, imports, and exports, are not available, although Glaeser et al. (1992) suggest this method as the most direct way of testing spillovers. Nevertheless, the second method of testing knowledge spillovers can still catch the general effect, albeit it cannot examine differences across industries. Therefore, in order to test knowledge spillovers in a feasible way, I adjust the second method of testing knowledge spillovers, and distinguish the impact of knowledge spillovers within industries, across industries, and from external sources on the development of new products respectively.

2.4.1 Data and Measurements

For the econometrical model, panel data is applied, and it encompasses 21 municipalities in Guangdong during the period 2001–2012. The scale of this geographical unit is more appropriate for knowledge spillovers to take place than that of the provincial unit. The starting year, 2001, is selected because it was during this year that all testing variables are available for a balanced panel dataset.

Dependent variable in the model is new product rate, i.e. share of new product in total industrial output value. In the research context of Guangdong province, where technological upgrading is characterized as improvements rather than radical innovations, new products emerge from improvements in technologies that are central to firms, the reverse engineering of new technological principles, and the recombination or redesign of existing components through the deep assimilation of existing technological principles (Kogut and Zander 1992). In spite of the inability to introduce new-to-the-world innovations, firms in Guangdong province have been intensifying R&D input almost to the level in OECD countries, as shown in Sect. 2.2. Therefore, it makes more sense to investigate the factors driving the innovation intensity rather than emergence of innovation.

The index of MAR externalities aims to measure the city's degree of specialization. As city-industry data are not applied in the model, location quotient can be calculated only for the most prominent industry in each city. In a panel dataset, however, location quotient has its limitation, inasmuch as that it assumes constant productivity and constant consumption pattern across all the cities, which is not realistic particularly in a dynamic term. Instead, the model introduces a more direct measure of specialization, which is the share of the city's biggest industry according to its output value. The specialization index in the model is defined as:

$$\text{Specialization index} = P_{\max}$$

where $P_{\max}$ is the share of the largest industry in the city according to output value.

Measuring a city's industrial diversity is a common way of testing the impact of Jacobs' externalities on the firm innovativeness. Indexes, such as Hirschman-Herfindahl Index (HHI) and entropy index, are mostly used in the literature. The difference between these two indexes is that HHI calculation is based on the square of the shares of respective industries, whereas entropy index calculation is based on the logarithm of the shares of respective industries. Thus, HHI tends to amplify the influence of stronger industries. The entropy index is chosen for the model because it is aimed to measure diversity instead of concentration.

$$\text{Diversification index} = \sum_{i=1}^{N} p_i \log\left(\frac{1}{p_i}\right)$$

where p_i is the share of industry i in the whole manufacturing output value.

Aside from measuring knowledge spillovers on the local scale, I also intend to assess the impact of external linkages on firm innovativeness in Guangdong. Here, import value and FDI are included in the model. The import value seeks to measure knowledge spillovers embedded in foreign commodities, whereas FDI aims to measure know-how spillovers from global production networks. The impact of overseas talents cannot be measured because of missing data. Two issues should be mentioned in the measurement of external knowledge spillovers. First, FDIs need time to become embedded in the local environment and to exert a knowledge spillover effect on local firms; therefore, in accordance with hypothesis two, FDI stock since 1985 is used instead of the annual FDI flow. Second, data on import value at the municipal level of is not categorized; thus, I cannot distinguish between original trade, processing and assembly trade, and compensation trade, among others. In fact, only a proportion of imports, such as imports for processing trade and imported equipment, can boost intra-firm learning processes, and ultimately increase the technological capabilities of firms. Therefore, the overall import value without subdivision is very likely to generate a biased estimation of the impact of import goods on firm innovativeness. Therefore, the result of import value should be carefully explained in the model.

To avoid the omitted variable bias in the model, two control variables are introduced.

1. Firm-level human capital: The management literature holds that technological upgrading and innovation is mainly a firm-level decision, and internal efforts are of major importance. Moreover, internal traits, particularly the level of human capital, determine to a large extent how firms search, absorb, and internalize knowledge spillovers and translate them into better technological performance. In this model, I use the number of personnel engaged in R&D activities to control for firm-level internal innovation efforts.

2. Urban externalities: Compared to dynamic externalities (knowledge spillovers) brought by MAR externalities and Jacobs' externalities, urban externalities bring two major advantages to firms located in larger cities in favor of technological upgrading and innovation. First, larger cities offer better infrastructure, especially transport infrastructure, such as highways and airports, which enhances the market access of firms. Second, larger cities offer larger odds of interaction (Glaeser 1999), exerting indirect influence on the impact of MAR externalities and Jacobs' externalities on technological upgrading and innovation. However, larger cities can also harm the local industry due to negative urban externalities, such as congestion, high factor costs, and pollution.

2.4.2 Static and Dynamic Approach for Panel Data

In panel dataset, a number of individuals are continuously observed over a period of time. Panel data helps in improving the estimation results in the following three aspects (Baltagi 2005): (a) panel data can control for individual heterogeneity, (b) they can catch the dynamics of adjustment compared to cross-sectional data; and (c) they have less collinearity among variables and they add more variability, including that within groups and between groups, compared to time-series and cross-section dataset.

The starting point of analysis is to estimate the static panel data regression models of innovativeness as a function of both local and non-local knowledge spillover. In static panel data model, there are basically two techniques, which utilize different scales of variation in the panel dataset. The fixed-effects model uses within-individual variations in the panel data, and the random-effects model makes use of both within-individual and between-individual variations. In the random-effects models, the absence of any correlation between the unit-specific residuals and the regressors is a required assumption. The fixed-effects model loses this limitation and assumes, in a realistic manner, that the unit-specific and often time-invariant characters is correlated with the regressors. In other words, the random-effects model is more efficient because it utilizes comprehensive information on the panel data, thereby catching both cross-section and within-individual data variations. However, it has a stricter assumption than the fixed-effects model. The Hausman test is then applied for evaluation on the effectiveness of the fixed-effects and random-effects models. If the null hypotheses of the Hausman test is rejected, namely with a significant value, the fixed-effects estimation is more appropriate than the random-effects. Otherwise, instrumental variables should be applied to deal with the endogenous problem, which leads to the use of the dynamic panel data estimation in the next step.

Dynamic approach for panel data, introduced by Arellano and Bond (1991), transforms the equation into first differences and uses lagged values of the endogenous variables as the instruments. Introduction of these lags is crucial to control for the dynamics of the process. Therefore, it is the most suitable techniques for testing the third hypotheses, in which regional innovativeness depends on its own past realization (path-dependent) and also gains from external knowledge spillover

Table 2.4 Descriptive statistics of the variables over time. (Data sources: Calculation based on data compiled from GPBS (2002, 2005, 2009, 2013))

	2001	2004	2008	2012
New product rate (unit: %)	4.6 (4.5)	8.0 (8.6)	7.0 (6.7)	10.8 (8.3)
Industrial Specialization (unit: %)	26.5 (15.4)	29.1 (17.3)	26.7 (14.4)	26.2 (12.6)
Industrial diversity	3.7 (0.59)	3.6 (0.67)	3.7 (0.56)	3.7 (0.47)
FDI stock (unit: USD billion)	6.2 (7.5)	8.2 (10.2)	11.2 (14.1)	15.2 (19.4)
Import value (unit: USD billion)	4.0 (7.5)	7.9 (16.2)	13.3 (28.0)	19.5 (43.9)
Firm personnel engaged in R&D activities (unit: Thousand)	2.7 (4.9)	3.4 (7.9)	9.5 (24.8)	24.7 (44.7)
Population (Unit: Million)	416.0 (211.9)	433.8 (212.9)	471.1 (250.9)	503.8 (286.5)

Numbers outside the brackets indicate average value; numbers in brackets indicate the standard deviation

cannot be realized in the short run. As many instruments would be generated in the dynamic model—with a number exceeding the independent variables—generalized method-of-moments (GMM) estimation techniques should be applied. Note that GMM estimates is only valid under the condition that no serial correlation exists. Accordingly, Arellano-Bond test is used as a diagnostic test for first and second order serial correlation.

2.4.3 Description of the Data

In Sect. 2.4.1, it has been explained in detail the construction of variables for estimation purposes. In this section, I make a general description of the data during the period 2001–2012 (Tabel 2.4). For the primary investigated variable—the new product rate, there has been a general uptrend in the mean value and the standard deviation, indicating an uneven growth across municipalities in Guangdong. As shown by Fig. 2.3, Shenzhen experiences the largest increase in new product development; the most dramatic increase occurred in 2008, and again during the recovery from the worldwide crisis in 2010. Other cities in the PRD, like Guangzhou and Foshan, also experience relatively large augment. Meanwhile, product development in Huizhou and Dongguan has a moderate growth, and other cities—mostly located in the periphery area of Guangdong province—have experienced fairly low growth.

The averages and standard deviations of the specialization and diversity index stayed quite stable over time, indicating a path-dependent pattern of industrial development. Cities tend to do the things they were good at in the past.

2.4 Model Specification and Data

Unit: billion Yuan

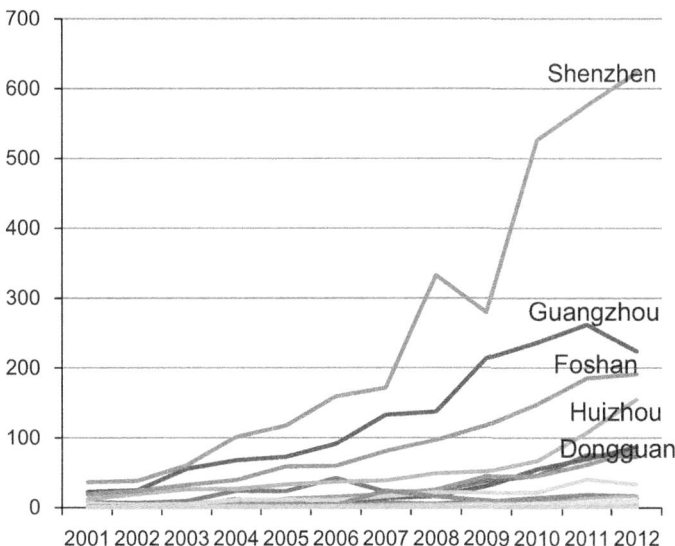

Fig. 2.3 New product output value over time among the municipalities. (Data source: GPBS (2002–2013, annual))

FDI stock increases more than twice in Guangdong in 2001–2012; however, the standard deviation also increases more than twice, showing the unbalanced flow and regional distribution of FDI in Guangdong. FDI tracks a path-dependent route because of the sinking cost of capital investment and increasing returns from the technological know-how of local firms developed in the process of transacting with foreign firms.

Import value follows a similar pattern as that of FDI stock across municipalities in Guangdong. In 2012, the import value in the core area of Guangdong, which includes the nine municipalities in the PRD, accounted for 97 % of the whole import value of the province. This may be attributed to higher consumption levels in these areas, leading to the importation of final goods, but more importantly to the prevalent mode of processing trade in the core region.

Finally, firm-level human capital also increases very unevenly across municipalities in Guangdong over time, whereas the population shows a comparatively stable pattern of distribution from 2001 to 2012. Large variations in most of the indicators justify the application of panel data to include more informative data in the model. As seen in the econometric results in the following Sect. 2.5, the dynamic differences among FDI stock, import value, and human capital, as well as their interaction effects with specialization and diversification, adequately explain the differences in new product development achievements across the 21 municipalities of Guangdong.

Table 2.5 Random effect of static innovation model for panel data. (Data sources: Calculation based on data compiled from GPBS (2002–2013, annual))

Independent variable	New product rate (Unit: %)					
	(1)	(2)	(3)	(4)	(5)	(6)
Specialization index (Spe.)	0.25*** (5.47)	–	0.28*** (5.04)	0.25*** (5.28)	–	–
Diversified index (Div.)	–	−5.70*** (−5.10)	–	–	−6.06*** (−4.38)	−5.55*** (−4.77)
FDI stock Unit: billion Yuan	0.42*** (3.86)	0.47*** (4.53)	0.48*** (3.86)	0.42*** (3.02)	0.29 (0.69)	0.51*** (3.99)
FDI stock*Spe.	–	–	−0.003 (−0.93)	–	–	–
FDI stock*Div.	–	–	–	–	0.04 (0.44)	–
Import Unit: billion Yuan	−0.09* (−1.84)	−0.10** (−2.04)	−0.05 (−0.87)	−0.09 (−0.68)	−0.08 (−1.32)	−0.02 (−0.12)
Import *Spe.	–	–	–	0.00 (0.01)	–	–
Import*Div.	–	–	–	–	–	−0.03 (−0.59)
R&D personnel Unit: thousand	0.10*** (2.72)	0.09*** (2.57)	0.10*** (2.78)	0.10*** (2.66)	0.09** (2.56)	0.09** (2.43)
Population Unit: Million	−0.014*** (−2.81)	−0.014*** (−3.08)	−0.014*** (−2.89)	−0.014*** (−2.79)	−0.014*** (−3.08)	−0.014*** (−2.99)
Number of Observations	252	252	252	252	252	252
Overall R^2	0.48	0.52	0.48	0.48	0.52	0.52
Hausman test	8.1	9.21	10.57	10.20	9,97	8.74

Standard errors in parentheses; *$p<0.10$, **$p<0.05$, ***$p<0.01$

2.5 Empirical Results

In both the static and dynamic model, the three hypotheses are to be consecutively examined. Table 2.5 presents the static panel data estimates for new product rate in Guangdong. The random-effects model is proved to be appropriate due to the insignificance of the Hausman test parameter, implying no correlation between the individually determined standard errors and the regressors.

As a baseline, I initially estimate the effects of different scales of knowledge spillovers without including the interaction term in model (1) and (2). Because the diversified index and specialization index are mutually substitutive and highly

2.5 Empirical Results

correlated, only one of the two regressors is included in the model at one time. Columns (1) and (2) show the results. As for knowledge spillovers on the local scale, whether they are measured by diversity (entropy index of the city's industries) or specialization (share of the city's largest industry), the result of our model confirms the first hypothesis: specialization contributes to product upgrading in Guangdong. In accordance with MAR externalities, knowledge spillovers within industries tend to generate a higher level of product innovation than knowledge spillovers across industries as suggested by Jacobs' externalities. In addition, as expected, FDI stock has a significantly positive impact on new product rate, whereas the impact of imports on new product rate is significantly negative. As shown by Tabel 2.1, imported goods in Guangdong is primarily comprised of electronics components. This result supports the second hypothesis by suggesting that imported goods may reduce the occurrence of knowledge spillover by weakening vertical inter-firm networks, especially when their absorptive ability is low for the positive mechanism of enhancing firm capabilities through learning by importing to take full effect. This view differs from other empirical studies confirming the positive impact of trade on productivity. However, considering the data and study context in this model, it is helpful to think about differences in trade impacts across the different developmental phases. Considering the positive impact of FDI stock, it can be concluded that the absorptive ability of Guangdong firms is high enough to avail themselves of knowledge spillovers brought by FDIs, but still not high enough to actively transform the knowledge spillovers brought by imported goods into their own technological capabilities. As expected by the traditional management literature, internal efforts are fairly important in new product development; an increase of 1,000 technological personnel increases the new product rate by over 10%. According to model (1) and (2), the disadvantages of urban externalities seem to outweigh their advantages at a significant level. The third hypothesis concerns how external knowledge spillovers interact with urban industrial structure in latecomer regions. To catch these interacting effects, I include the interaction term separately in models (3) and (6) to avoid correlation problems. However, no statistically significant effect is shown for all the four interaction terms in the models.

The ineffectiveness of the third hypothesis in the static model might lie in the existence of dynamic effects. After controlling for the dynamic process, new or different relationships between the dependent and independent variables are likely to emerge. The use of dynamic panel data techniques in Tabel 2.6 allows to uncover the moderating effect of urban industrial structure on the impact of external knowledge spillover on innovation. Test for serial correlation confirms the presence of first order correlation, which is required in the differenced estimates, but no second order autocorrelation. Therefore, the GMM estimators are valid.

Similar to Tabel 2.5, the first two models are run without the interaction term as the baseline estimations. The coefficients for specialization and diversification corresponds with those in the static estimations, yet not in a significant level. Results for the FDI stock and import values, on the other hand, reveal a richer picture. FDI stock in the present year actually curtails the firm innovativeness on the regional scale, whereas FDI stock in the last year is finally able to boost innovation. This

Table 2.6 GMM estimates of a dynamic innovation model for panel data. (Data sources: Calculation based on data compiled from GPBS (2002–2013, annual))

Independent Variable	New product rate (Unit: %)			
	(1)	(2)	(3)	(4)
Dependent lag 1	0.42***	0.44***	0.45***	0.46***
	(8.15)	(7.36)	(7.17)	(8.15)
Specialization index (Spe.)	0.32	–	–	–
	(1.57)			
Diversified index (Div.)	–	−8.78	−9.91	−7.97
		(−1.60)	(−1.31)	(−1.39)
FDI stock Unit: billion Yuan	−1.79**	−1.72**	−3.55	−2.06**
	(−2.06)	(−2.12)	(−1.53)	(−2.21)
FDI stock lag 1	2.30**	2.06**	2.01*	2.26**
	(2.34)	(2.28)	(1.90)	(2.49)
FDI stock*Div.	–	–	−0.03	–
			(−0.07)	
FDI stock*Div. lag 1	–	–	0.53***	–
			(2.94)	
Import Unit: billion Yuan	0.003	0.018	0.09	0.54**
	(0.04)	(0.31)	(1.42)	(2.22)
Import lag 1	−0.18***	−0.17***	−0.13***	−0.70***
	(−3.20)	(−3.19)	(−2.72)	(−2.63)
Import*Div.	–	–	–	−0.15*
				(−1.83)
Import*Div. lag 1	–	–	–	0.17*
				(1.81)
R&D personnel Unit: thousand	0.15***	0.15***	0.16***	0.14***
	(3.26)	(3.25)	(2.66)	(2.80)
Population Unit: Million	−0.019	−0.010	−0.014	−0.001
	(−1.22)	(−0.54)	(−0.93)	(−0.06)
Number of observations	210	210	210	210
AR(1)	−2.43**	−2.33**	−2.42**	−2.22**
AR(2)	1.65	1.75*	1.53	1.45

Standard errors in parentheses; *$p<0.10$, **$p<0.05$, ***$p<0.01$

result can be partly explained by the fact that FDI flows in the present year are mostly aimed for production expansion, diluting the total impact of FDI stock on firm innovation. As the foreign invested firms become more embedded in the local production networks—even the foreign firms are not undertaking innovation themselves—they tend to disseminate advanced technological know-how to local firms. Likewise, the negative effect of import flow, as reported by the static model, just started to exert its influence on regional innovativeness with 1 year delay. The domestic firms are able reap innovative gains promptly from the advanced imported components at the very start, as suggested by the positive coefficient in the present year, but failed to make sustainable innovative outcomes owing to low absorptive capacity to exploit more in the long run. This result corresponds with the second hypothesis. As a result, the local firms tend to rely on imported components instead of making efforts in testing and adapting. Overall, the introduction of hysteretic effect demonstrates the dynamic feature of the external knowledge spillover.

In model (3) and (4), the interaction terms between FDI stock and import and regional diversification are included separately. In fact, the interaction with regional specialization is also tested out. However, the econometrical results for the latter are not significant, so it is not include in Tabel 2.6. Estimation results from model (3) show that the first lag for the interaction term between FDI stock and diversification is positive at 1 % significant level, although the effect of FDI stock alone has been weakened compared to the estimations without the interaction. The estimates corroborates the third hypothesis, implying that the impact of FDI stock on new product development relies on the urban industrial structure—the more diversified the firms in the region produce, the larger is the impact that FDI brings into play.

In model (4), the interaction term between import and diversification index is a little bit complex but interesting. For the current level of import, previous deduction that local firms make efforts in learning and incremental innovation is corroborated with a more significant and larger coefficient. It shows, however, that the diversified economy tend to curtail the firms' efforts to learn from the imported components in the beginning, which is related the lack of specialized know-how in the local economy. One the other side, results of the first order of time lag for import alone are in line with the previous three models, with a stronger and more significant effect, implying that the impact of import on facilitating innovation is not sustainable in the long run. Furthermore, the interaction term suggests that urban diversified economy is able to offset in some way the negative effect of import in the long run when absorptive capacity is low, and contribute to the regional innovativeness with diverse knowledge stock that stimulates innovations through new application and combination of the advanced components in other related product markets. Generally speaking, the estimation is consistent with the result of model (3), which also implies that the degree of urban economic diversification determines to which extent import trade influences on new product development.

2.6 Discussion and Conclusion

The results of this chapter shed light on the role of knowledge spillovers in regional innovation of latecomer countries in the era of globalization. Corresponding to Neffke et al. (2011) argument on the relationship between the industrial cycle and knowledge spillovers on the local scale, empirical investigation in the chapter shows that knowledge spillovers within industries matter more when technological activities are incremental and are oriented towards improvements. More importantly, the analysis considers external knowledge spillovers, which are suggested by Kuznets (1973) as "advantages of backwardness", as the key trigger for localized knowledge spillovers within industries through the mechanism of FDIs and import.

However, external knowledge spillovers do not take place right away. In fact, the long-term effect of external spillovers are closely related to the investment stock, degree of embeddedness, and absorptive ability of local firms. When the absorptive ability of firms is low, reliance on imported components actually harm the localized knowledge spillover mechanism by reducing the intensity of inter-firm linkages. The econometrical practices further find out the moderating role of diversified local economy on the triggering effect of external knowledge spillover in the long run. All these results have deepened our understanding on the complex interaction of knowledge spillover mechanism on both global and local scale.

From the empirical results, the importance of enhancing the absorptive capacity of local firms should be highlighted as the key policy implication. In the post-crisis era, Guangdong government in both the provincial and municipal level should devote efforts into attracting talent and providing incentives for small and medium-sized firms to invest in technological upgrading, with a final aim to exploit the inflow of external knowledge in such an out-oriented regional economy. Since the onset of the financial crisis, China has faced great pressures from renminbi (RMB) appreciation. In order to rebalance the trade surplus and stabilize the exchange rate, import volume, particularly the volume of high-tech products, is expected undergo constant increase. Imports can be utilized in a positive way for innovation without harming the local density of networks—only by enhancing the technological capabilities of local firms. This strategy is particularly important for the Chinese firms when the export market is shrinking.

The chapter imply that external knowledge spillovers are keys to the catch-up efforts of latecomer countries and the conditional convergence between technological leaders and technological followers. However, as indicated by Helpman (1993), and also Barro and Sala-i-Martin (1992), the low cost of imitation and the absence of intellectual property rights give technological followers the incentive to copy as a means of achieving growth. However, as the technological gap decreases and the cost of imitation increases, the growth rate of technological followers tends to decrease. This line of thought provide some hints for future research directions. First, convergence does not take place as long as a technological gap between firms exists. Convergence requires firms to avail themselves of external knowledge spillovers and convert these spillovers into their own innovative capabilities. Therefore, firm attributes and organizations should come into the focus of analysis. Although the aggregate analysis in this chapter can provide informative implications on the impact of external knowledge spillovers within the latecomer context, the channels through which innovation in latecomer regions work are still far from clear. As suggested by Martin and Sunley (2007), economic change and system adaptation is complex and differentiated across space. In other words, the traits of firms and regions should be paid attention to when exploring the success of latecomer regions in availing themselves of external knowledge spillovers (Budd and Hirmis 2004). Second, in the absence of a framework for the protection of intellectual property rights in transition economies, informality becomes an important aspect to consider in the future study.

Overall, this study calls upon econometric studies that take a dynamic perspective and procedure towards innovation and knowledge spillovers. Moreover, it suggests that a firm-level study must go beyond the general information provided by econometric studies, and further explore the microeconomic aspects of the technological upgrading activities of latecomer countries and regions.

References

Amin A (2000) Industrial districts. Blackwell, Oxford
Arellano M, Bond S (1991) Some tests of specification for panel data: monte Carlo evidence and an application to employment equations. Rev Econ Stud 58(2):277–297
Arrow KJ (1962) The economic implications of learning by doing. Rev Econ Stud 29(80):155–173
Asheim BT (2000) Industrial districts: the contributions of marshall and beyond. Oxford University Press, Oxford
Baltagi BH (2005) Econometric analysis of panel data, 3rd edn. Wiley, Chichester
Barro RJ, Sala-i-Martin X (1992) Convergence. J Political Econ 100(2):223–251
Bianchi G (1998) Requiem for the third Italy? Rise and fall of a too successful concept. Entrep Reg Dev 10(2):93–116
BMBS (Beijing Municipal Bureau of Statistics) (2013) Beijing Tongji Nianjian (Beijing Statistical Yearbook). China Statistics Press, Beijing
Boschma R (2004) Competitiveness of regions from an evolutionary perspective. Reg Stud 38(9):1001–1014
Boschma R, Lambooy JG (2002) Knowledge, market structure, and economic coordination: dynamics of industrial districts. Growth Chang 33(3):291–311
Boschma R, Iammarino S (2009) Related variety, trade linkages, and regional growth in Italy. Econ Geogr 85(3):289–311
Branstetter L (2001) Are knowledge spillovers international or intranational in scope? Microeconometric evidence from the US and Japan. J Int Econ 53(1):53–79
Branstetter L (2006) Is foreign direct investment a channel of knowledge spillovers? Evidence from Japan's FDI in the United States. J Int Econ 68(2):325–344
Budd L, Hirmis AK (2004) Conceptual framework for regional competitiveness. Reg Stud 38(9):1015–1028
Carlton DW (1983) The location and employment choices of new firms: an econometric model with discrete and continuous endogenous variables. Rev Econ and Stat 65(3):440–449
Chang HJ (1993) The political-economy of industrial-policy in Korea. Camb J Econ 17(2):131–157
Cheung K, Lin P (2004) Spillover effects of FDI on innovation in China: evidence from the provincial data. China Econ Rev 15(1):25–44
Coe DT, Helpman E, Hoffmaister AW (1997) North-south R&D spillovers. Econ J 107(440):134–149
Cohen WM, Levinthal DA (1990) Absorptive capacity: a new perspective on learning and innovation. Adm Sci Q 35(1):128–152
Combes PP (2000) Economic structure and local growth: France, 1984–1993. J Urban Econ 47(3):329–355
CSSB (China State Statistical Bureau) (2013) Zhongguo Tongji Nianjian (China Statistical Yearbook). China Statistics Press, Beijing
Falvey R, Foster N, Greenaway D (2004) Imports, exports, knowledge spillovers and growth. Econ Lett 85(2):209–213
Feldman MP (1994) The geography of innovation. Kluwer, Boston

Feldman MP (2000) Location and innovation: the new economic geography of innovation, spillovers, and agglomeration. Oxford University Press, Oxford
Feldman MP, Audretsch DB (1999) Innovation in cities: science-based diversity, specialization and localized competition. European Econ Rev 43(2):409–429
Giuliani E (2005) Cluster absorptive capacity: why do some clusters forge ahead and others lag behind? European Urban and Reg Stud 12(3):269–288
Glaeser EL (1999) Learning in cities. J Urban Econ 46(2):254–277
Glaeser EL, Kallal HD, Scheinkman JA et al (1992) Growth in cities. J Political Econ 100(6):1126–1152
Gort M, Klepper S (1982) Time paths in the diffusion of product innovations. Econ J 92(367):630–653
GPBS (Guangdong Provincial Bureau of Statistics) (1998, 2001–2013, annual) Guangdong Tongji Nianjian (Guangdong Statistical Yearbook). China Statistics Press, Beijing
Griliches Z (1990) Patent Statistics as economic indicators: a survey. J Econ Lit 28(4):1661–1707
Grossman GM, Helpman E (1990) Trade, innovation, and growth. Am Econ Rev 80(2):86–91
Helpman E (1993) Innovation, imitation, and intellectual property-rights. Econometrica 61(6):1247–1280
Henderson JV (2003) Marshall's scale economies. J Urban Econ 53(1):1–28
Henderson R, Cockburn I (1996) Scale, scope, and spillovers: the determinants of research productivity in drug discovery. Rand J Econ 27(1):32–59
Hoover EM (1937) Location theory and the shoe leather industries. Harvard University Press, Cambridge
Humphrey J, Schmitz H (2004) Chain governance and upgrading: taking stock. In: Schmitz H (ed) Local enterprises in the global economy: issues of governance and upgrading. Edward Elgar, Cheltenham
Ivarsson I, Alvstam CG (2005) The effect of spatial proximity on technology transfer from TNCs to local suppliers in developing countries: the case of AB Volvo in Asia and Latin America. Econ Geogr 81(1):83–111
Jacobs J (1969) The economy of cities. Random House, New York
Jaffe AB (1989) Real effects of academic research. Am Econ Rev 79(5):957–970
Jaffe AB, Trajtenberg M, Henderson R (1993) Geographic localization of knowledge spillovers as evidenced by patent citations. Q J Econ 108(3):577–598
Javorcik BS (2004) Does foreign direct investment increase the productivity of domestic firms? In search of spillovers through backward linkages. Am Econ Rev 94(3):605–627
JPBS (Jiangsu Provincial Bureau of Statistics) (2013) Jiangsu Tongji Nianjian (Jiangsu Statistical Yearbook). China Statistics Press, Beijing
Kleinknecht A, Van Montfort K, Brouwer E (2002) The non-trivial choice between innovation indicators. Econ Innov New Technol 11(2):109–121
Kogut B, Zander U (1992) Knowledge of the firm, combinative capabilities, and the replication of technology. Organ Sci 3(3):383–397
Kroll H, Tagscherer U (2009) Chinese regional innovation system in times of crisis: the case of guangdong. Asian J Technol Innov 17(2):101–128
Krugman P (1991) History and industry location—the case of the manufacturing belt. Am Econ Rev 81(2):80–83
Kuznets S (1973) Modern economic growth: findings and reflections. Am Econ Rev 63(3):247–258
Lemoine F, Unal-Kesenci D (2004) Assembly trade and technology transfer: the case of China. World Dev 32(5):829–850
Loury GC (1979) Market structure and innovation. Q J Econ 93(3):395–410
Lucas RE (1988) On the mechanics of economic development. J Monet Econ 22(1):3–42
Marshall A (1920) Principles of economics. Macmillan, London.
Martin R, Sunley P (2007) Complexity thinking and evolutionary economic geography. J Econ Geogr 7(5):573–601

References

Maskell P, Malmberg A (2007) Myopia, knowledge development and cluster evolution. J Econ Geogr 7(5): 603–618

Miraky W (1994) The firm product cycle. MIT Mimeograph, Cambridge.

Neffke F, Henning MS, Boschma R et al (2011) The dynamics of agglomeration externalities along the life cycles of industries. Reg Stud 45(1):49–65

Organization for Economic Co-operation and Development (OECD) (2008) Main Science and Technology Indicators database. OECD, Paris

Oro K, Pritchard B (2011) The evolution of global value chains: displacement of captive upstream investment in the Australia-Japan beef trade. J Econ Geogr 11(4):709–729

Owen-Smith J, Powell WW (2004) Knowledge networks as channels and conduits: the effects of spillovers in the Boston biotechnology community. Organ Sci 15(1):5–21

Parrado R, De Cian E (2014) Technology spillovers embodied in international trade: intertemporal, regional and sectoral effects in a global CGE framework. Energy Econ 41:76–89

Parthasarathy B, Aoyama Y (2006) From software services to R&D services: local entrepreneurship in the software industry in Bangalore, India. Environ Plan A 38(7):1269–1285

Peng MW, Wang DYL, Jiang Y (2008) An institution-based view of international business strategy: a focus on emerging economies. J Int Bus Stud 39(5):920–936

Power D, Lundmark M (2004) Working through knowledge pools: labour market dynamics, the transference of knowledge and ideas, and industrial clusters. Urban Stud 41(5–6):1025–1044

Revilla Diez J Kiese M (2006) Scaling innovation in South East Asia: empirical evidence from Singapore, Penang (Malaysia) and Bangkok. Reg Stud 40(9):1005–1023

Romer PM (1986) Increasing returns and long-run growth. J Polit Econ 94(5):1002–1037

Rosenthal SS, Strange WC (2003) Geography, industrial organization, and agglomeration. Rev Econ Stat 85(2):377–393

Saxenian A. (1994) Regional advantage culture and competition in silicon valley and route 128. Harvard University Press, Cambridge

Saxenian A, Hsu JY (2001) The silicon valley–Hsinchu connection: technical communities and industrial upgrading. Ind Corp Change 10(4):893–920

SMBS (Shanghai Municipal Bureau of Statistics) (2013) Shanghai Tongji Nianjian (Shanghai Statistical Yearbook). China Statistics Press, Beijing

Storper M (1995) The resurgence of regional economies, ten years later: the region as a nexus of untraded interdependencies. Eur Urban Reg Stud 2(3):191–221

Trajtenberg M (1990) A penny for your quotes—patent citations and the value of innovations. Rand J Econ 21(1):172–187

Wang X (2008) Guangdongsheng gongye qiye chuangxindiaocha fenxi (Analysis of Guangdong industrial enterprise innovation survey). Zhujiang jingji (South China Review) 6:55–62

Wang CC, Lin GCS (2013) Dynamics of innovation in a globalizing china: regional environment, inter-firm relations and firm attributes. J Econ Geogr 13(3):397–418

Wei YD, Liu Y, Chen W (2009) Globalizing Regional Development in Sunan, China: does suzhou industrial park Fit a Neo-Marshallian district model? Reg Stud 43(3): 409–427

Yeung HWC (2009) Regional development and the competitive dynamics of global production networks: an east asian perspective. Reg Stud 43(3):325–351

ZPBS (Zhejiang Provincial Bureau of Statistics) (2013) Zhejiang Tongji Nianjian (Zhejiang Statistical Yearbook). China Statistics Press, Beijing

Chapter 3
Interactive Learning and Systemic Innovation

Firm-Level Evidence from the Electronics Industry in the Pearl River Delta, China

Abstract In this chapter, I aim to investigate the role of interactive learning in promoting firm innovation. By investigating the willingness and capacity of firms to undertake interactive learning in product innovation, this chapter sheds light on the emergence of dynamic externalities of a regional innovation system at the firm level. Based on a survey on innovative electronics firms in the PRD, China, this chapter demonstrates that firms undertaking the highest intensity of interactive learning with the widest scope of business partners, such as foreign customers, domestic customers, parent companies, universities and sales agents, tend to achieve better innovation outcomes. What's more, the intensive interactive learning firms have a much higher tendency to apply informal *Guanxi* networks, for example with business partners, relatives and friends, than other firms. Overall, this chapter contributes to the understanding of the form and effect of interactive learning in the Chinese context. Finally, the chapter addresses the possible lock-in issue and points out future research questions on the changing pattern of interactive learning with a maturing institutional framework in transition China.

3.1 Introduction

In an industrial cluster, the inter-firm linkage, which are oriented towards the reduction of transaction costs, creates static externalities among the clustering firms (Marshall 1920; Williamson 1981). In addition, it also induces the dynamic externalities for increasing returns in the whole economy, where one input into innovation activities in a single firm is able to generate disproportionately more output in the whole cluster owing to spillover effects based on the interactive learning process. The interactive learning is quasi-cooperative relationships, and can be undertaken among groups of users and producers as well as between business firms and knowledge-producing institutes (Cooke et al. 1997; Howells 1999; Revilla Diez 2000; Asheim and Coenen 2005; Asheim and Parrilli 2012). Overall, the

---

This chapter has been revised considerably and published as a second edition in *Interactive learning, informal networks and Innovation: Evidence from electronics firm survey in the Pearl River Delta, China*. Fu W, Revilla Diez J, Schiller D. Research Policy 42/3, Copyright © 2013, Elsevier.

willingness and capacity of clustering firms to undertake interactive learning is vital for long-term economic growth dynamics (Arrow 1962; Romer 1986; Krugman 1991; Cooke et al. 1998).

Ever since the beginning of the opening policy in 1978, industrial clusters have been emerging in coastal areas of China, taking the opportunity of relocation and subcontracting processing functions from global lead firms. Firms in these agglomerations draw significantly on the static externalities advantages, such as saving intermediate goods costs and sharing supported infrastructure. Moreover, flexible and responsive production is able to be sustained thanks to the use of networked enforcement mechanisms based on informality and cooperation (Meyer et al. 2009). In this way, the specialization of industrial clustering in China facilitates the use of comparative advantages such as cheap labor and land.

Nevertheless, the source of regional economic growth does not lie in static externalities, but in the dynamic externalities that generated by interactive learning and innovation synergies between the economic agents to jointly exploit the new combinations and market opportunities, creating increasing return to the large stock of knowledge in the clusters. For the regional innovation systems in China, the formation of this dynamic mechanism is of particular importance as the dependence on FDI for technology persisted for long which contrasted with the poor interaction among the local firms. It is expected that interactive learning should take place as the local firms have accumulated certain level of balanced knowledge stock, enabling them to reciprocally learn from each other. Even when the high-end technology might still originate from global lead firms, the local firms can be motivated by the imported technology with the joint-exploitation of new market opportunities and combinations through interactive learning. Hence, the aim of this chapter is to find out how interactive learning is strategically managed and how the scope and intensity of interactive learning contribute to innovation outcome.

This study focuses on one of the largest electronics industry clusters in the PRD, China. The issue of interactive learning and systematic innovation is very relevant in the electronics industry in China. Firstly, Chinese electronics firms are mainly technology adopters, integrating "off the shelf" subcomponents into new product design. The modularity in the electronics industry necessitates the interaction between specialized firms to exploit the new market opportunities of new combinations, as well as to solve the technical problems of integrating components. Secondly, the technological frontier is moving at an astounding rate in the electronics industry. As indicated by Gjerde et al. (2002), firms are forced to innovate out of fear in missing innovation opportunities when the technological frontier pushes forward quickly. Therefore, by investigating the willingness and capacity of electronics firms in the PRD, China to undertake interactive learning in product innovation activities, this chapter sheds light on whether dynamic knowledge spillover externalities emerge in China. In the face of the global recession and domestic inflation, the capacity to draw on innovation externalities is of great importance for regional structural adjustments and long-term development.

The remainder of this chapter is structured as follows. Section 3.2 elucidates the need and the scope of undertaking interactive activities in order to promote innovation. Hypotheses are derived based on the theoretical discussion. Section 3.3 presents the dataset, key variables and the methodology. Section 3.4 discusses the empirical results. Finally, Sect. 3.5 concludes and discusses policy implications.

3.2 Innovation as an Interactive Process

Unlike such exogenous inputs as capital and labor, innovation and learning contribute to the improvement of productivity and determine the long-term economic growth (Arrow 1962; Romer 1986; Nelson and Siegel 1987). March (1991) emphasizes the importance of knowledge diversity in learning process and defines the dichotomy between exploration and exploitation. While exploitation makes exclusive use of the existing knowledge, exploration enables the firms to make use of new opportunity and avoid competency traps. Hence, the firms need to go beyond the organizational boundary and interact with external agents in order to undertake knowledge exploration. In Lundvall (1992)'s seminal work on national systems of innovation, he proposed that the approach towards systemic innovation and interactive learning considers the stock and rate of the R&D investment as the new determining variable in economic growth. That is to say, interactive learning creates increasing returns for the stock of knowledge and thus underpins long-term economic growth.

In order to tap into the interactive learning activities, it is useful to discuss the nature of knowledge. Salter and Reddaway (1969) distinguish between different types of production-relevant knowledge: firm-specific knowledge, sector product-field specific knowledge and generally applicable knowledge. Asheim and Coenen (2005) further elaborate the dimension of sector product-field specific knowledge into synthetic knowledge and analytical knowledge. The synthetic knowledge is based on know-how and experience, and contributes to the strengthening of industrial specialization. The analytical knowledge, on the other hand, is more general in a scientific sense, requiring frequent industry-university interaction and cooperation.

Firm-Specific Knowledge The firm-specific knowledge is well elaborated on by Nelson and Winter (1982)'s proposition of organizational routine. Routine consists of particular resources, skills, experience and know-how that the firm accumulates over time (Levitt and March 1988), and is therefore difficult to imitate for others.

Firm-specific knowledge are accumulated over time and selected by competitive market, determining the competitiveness of the firms (Teece et al. 1997). More importantly, organizational routines develop in a path-dependent manner, in which the firm tends to search for information and undertake activities related to its own knowledge sphere (Kline and Rosenberg 1986). Therefore, the firm displays bounded rationality and competence in the innovation-related activities, which has two important implications for the role of interactive learning in innovation.

Firstly, bounded rationality implies that the decision making process is determined by limited information, limited knowledge and limited resources of the individuals or entities, and thus leading them to base decision making on existing knowledge and capacity, which results in a satisfactory solution rather than an optimal one based on total rationality (Simon 1957; Simon 1991). Firms tend to identify the knowledge that is similar to their knowledge stock among the bulk of information, leading to larger chance of similar knowledge selected by firms for further learning. In another way, the ability to search relevant information for innovation is limited, which necessitates interactive learning to expand the scope of knowledge and competence. As a result, firms with bounded rationality are not able to calculate the result of decision-making on innovation investment when faced with uncertainty in the environment. In order to reduce risk-related uncertainty, firms have to collect more technical information and market information from outside organizations.

Secondly, the firms are normally not versatile kings but with bounded competence in a limited range of products and processes. As a result, firms are constantly confronted with technical problems in the innovation process which lie outside their range of knowledge and competence (Smith 2000). This kind of knowledge is not only limited to codified knowledge, such as the support of specialized equipment and operating software, but also refers to the more important tacit knowledge such as technical know-how and experience, which is key to problem-solving in the process of prototype development and technical design. Due to the tacitness of most knowledge, the firms need to engage in face-to-face interaction with other organizations in order to solve these problems and optimize the innovation outcomes.

In a broader sense, the market selection process reshuffles the relative efficiency of competing firms, and impels the firms to constantly monitor and adapt by developing new products and new markets, applying new set of inputs and new processes, and thus finally to adjust the organizational routine, which are all mainly based on the current knowledge and capacity (Nelson and Winter 1982). This process is more like a swirling ladder than a linear growth process. The adjustment process might be stagnated or slowed down because of the limited knowledge and capacity. It might be also backward because the firms might make mistakes. Whatsoever, the adjustment process, which underlies its dynamic competitiveness in the market, necessitates knowledge transfer within and between firms and constant learning due to the newness of knowledge for the firms.

Based on the above discussion, I propose the following proposition, upon which the logic of the empirical analysis is established:

Proposition 1 A: Due to bounded rationality and competence, firms need to complement internal efforts in innovation with interaction with other organizations in order to facilitate innovation-related decisions by searching for relevant information, and to support innovation implementation with external codified and tacit knowledge.

Sector Product-Field Specific Knowledge Tacit knowledge is not only confined to individuals or groups of co-operating individuals, but is also embedded within specific industries, which is often referred to in the literature as the "technological paradigm" (Dosi 1988). Technological paradigm refers to the common technological features, such as technical parameters, performance characteristics, and use of materials shared by firms in a specific industry (Smith 2000). Moreover, sector product-field specific knowledge also covers knowledge on markets, such as customer needs and the supply of industry-specific skills.

Therefore, firms within the same production field are close in cognitive proximity, which facilities the interactive learning process (Boschma 2004). Cognitive proximity within the same industrial space and supplier link would affect the search and imitation costs when exploiting knowledge. North (1996) proposes that social capital would affect the vertical division of labor. Futhermore, Lundvall (2005) argues that learning by interacting between the users and producers is able to make most use out of the learning by doing and learning by using effect within the organizational boundary, creating knowledge embodied in new machinery, new components and new software-systems.

Kline and Rosenberg's (1986) early work on the "chain-linked model of innovation" suggests that increased demand of the user firms would generate rapid rate of technical changes for the suppliers. Specifically, in the chain of innovation from the initial design to the production process, the later phase shifts more towards systematic interaction with user needs. In the Aalborg school of innovation systems, innovative activities within the vertically organized units have been the analytical focus. The search strategies and learning processes organized within the prevalent vertical linkages between the firms and their supplying firms of intermediate and capital goods distribute and transmit the qualitative knowledge related to product innovation (Lundvall 1988; Lundvall et al. 2002). In order to secure profitable innovation outcomes, the user-producer interaction must be in place to ensure constant feedback on needs, adjusted design, and performance (Hage and Alter 1997). Asheim and Gertler (2005) further elaborate that interactive learning between users and producers often takes place in industries in which synthetic knowledge is dominant. Synthetic knowledge pertains to the importance of applied and problem-solving knowledge, where the innovation process is oriented towards new combinations, new solutions and new utility concerning the user demands.

Interactive processes of knowledge transfer within supplier linkages bring about dynamic synergies rather than static efficiency on transaction cost reduction (Capello 1999). In the dynamic synergies between customers and suppliers, market information is constantly exchanged, while experience and know-how are shared through engineering knowledge instruction and quality monitoring (often undertaken by the customers). Consequently, the technology trajectory is co-evolving due to the coordination of the production process. In the context of latecomer countries, the firms also rely heavily on the parent companies and foreign customers to acquire advanced codified knowledge and also to absorb the codified knowledge with the engineers and managers sent by foreign partners for instruction on site (Morrison et al. 2008; Yang 2009; Yeung 2009).

In addition to vertical collaboration, innovative cooperation among horizontal firms also plays a role in the aspect of exchanging sector-specific know-how. Teece (1986) argues that when new products can easily be reverse engineered and imitated by competitors, firms tend to establish partnerships or alliances with other firms who have the potential capacity to produce them.

The following proposition provides a brief summary of how interactive learning is undertaken within the scope of sector product-field specific knowledge, which the empirical investigation will take into account.

Proposition 1B: Interactive learning within the vertically linked agents, i.e. between suppliers and customers, ensures the effective exchange of market information and constant feedback on technical problems and product adjustment, and thus promotes the product innovation outcomes.

Generally Applicable Knowledge Generally applicable knowledge refers to applicable knowledge in a broader sense. This generic knowledge is more about the scientific "know-why" knowledge that is playing an increasingly important role in the problem-solving of innovative efforts (Lundvall and Johnson 1994). It is of greater relevance for high-tech industries such as electronics, pharmaceuticals and chemistry, where the technological frontier is pushing forward at a rapid rate.

In contrast to the synthetic knowledge, which is more connected to sector product-field specific knowledge, Asheim and Gertler (2005) conclude that analytical knowledge is dominated by scientific know-why knowledge and is generated from internal documentation activities as well as collaboration with research institutes. From the research on Danish clusters, Jensen et al. (2007) also found that the mode of learning by doing, using and interacting is no longer able to sustain the competitiveness of firms. Firms that combine the DUI (doing, using and interacting) mode with the STI (science, technology and innovation) mode, i.e. connecting systematically with sources of codified and scientific knowledge, outperform other firms in terms of finding new solutions and developing new products. Systematic connection with generic scientific knowledge can be achieved in the following two ways.

Firstly, generic scientific knowledge can be absorbed through internal efforts such as R&D activities, reverse engineering and licensing into tacit knowledge. R&D function often exists in large firms due to the scale economy of research activities. Actually, the role of R&D activities in generating new knowledge in the context of latecomer firms is insignificant. More importantly, R&D activities display a social rate of return by influencing the absorptive capacity of the firms (Griffith et al. 2003), determining the capability of firms to transform externally codified scientific knowledge into their own routines of more tacit knowledge. For latecomer firms in particular, they can also gain access to advanced codified knowledge either through reverse engineering of the import products from global lead firms, or through formal licensing of the codified knowledge such as patents. However, the efficiency of these activities is determined by the absorptive capacity of firms to adapt them to their own specific needs.

Secondly, interaction with universities and research institutes assists firms in acquiring new knowledge through their intra- and interregional networks as well as in applying abstract scientific knowledge to operation and production. Generally applicable knowledge cannot be immediately applied to commercial needs and the spillover risk for the knowledge investors is too high. The public sector such as universities and research institutes, which normally operates without

3.2 Innovation as an Interactive Process

profit-maximization goals, is then involved in the production of generally applicable knowledge due to the problem of appropriability (Smith 2000). Other than distribution of knowledge between suppliers and customers, the distribution of knowledge among universities, research institutes and industry is also one of the most important aspects relevant for innovation activities (David and Foray 1995).

Overall, the following two proposition are proposed:

Proposition 1 C: Use of external scientific knowledge depends on the firms' absorptive capacity, which is accumulated by such activities as R&D activities, reverse engineering and patent licensing.

Proposition 1D: Interactive learning with universities and research institutes assists firms in acquiring new knowledge through their intra- and interregional networks as well as in applying abstract scientific knowledge to their own production needs.

Based on proposition 1 A-1D, it can be concluded that interactive learning is needed both in the decision-making and implementation processes of innovation due to the bounded rationality and bounded competence of firms, and it extends the scope of supplier linkages to knowledge-generating institutes. The scope of interactive learning actually brings superadditivity among its effects, which Storper and Venables (2004) refer to as "buzz effect". The superadditivity refers specifically to the increasing return of knowledge stock, in which the more information and knowledge the firms acquire through a wide scope of interaction, the more easily they are able to understand complex ideas. Moreover, Lundvall (2005) also implicitly implies that the scope of interactive learning should be widened in order to escape the lock-in effect of learning with limited numbers of organizations, especially in a sector with turbulent technological change and rapid change in customer needs. Therefore, firms in a buzz environment are highly motivated to undertake more complex innovation activities that are more likely to produce significant innovation outcomes.

Furthermore, firms need not only to extend the scope of interactive learning, but also to intensify interaction in order to better absorb each piece of information and knowledge from external organizations. The more the innovation is distinct from firms' existing knowledge and competence, the more efforts they should put into unlearning old routines and learning new ones. Due to the path-dependent accumulation of knowledge development, as previously discussed, firms have to undertake the interactive learning activities to an intensified degree, because they tend to return to their old ways of cognition and practice in the interaction process. Moreover, new codes have to be developed on a trial and error basis in innovation activities, especially the ones with higher rate (Lundvall 1992; Meeus et al. 2001). Therefore, firms must intensify the interaction with customers and other knowledge-producing institutes in order to ensure their success in developing new products.

Finally, the general hypothesis is advocated and its validity is to be tested in the empirical analysis:

Hypothesis: The scope and intensity of interactive learning with customers, universities and research institutes in the innovation process contribute to the innovation outcomes.

3.3 Survey Data and Methodology

The data applied in the following analysis is a set of standardized questionnaire data on electronics firms in the PRD, China, which was collected during the period between September and November 2009. The company survey was aimed at electronics firms mainly in four cities on the eastern coast of the PRD, where the electronics industry is dominant (as in Shenzhen and Dongguan) or developing very quickly (as in Huizhou and Heyuan). The questionnaire survey was conducted via telephone and mail, in which the questions were addressed to the Chief Executive Officers (CEOs) or senior executives of electronics companies in the PRD. Follow-up was conducted via telephone, aiming to complete the unanswered questions and improve the quality of the questionnaires. In total, 793 firms were contacted and 422 firms filled out the questionnaires, and the response rate is 53%. Among the surveyed firms, 167 are located in Shenzhen, 177 in Dongguan, 67 in Huizhou and 11 in Heyuan. In the analysis, I concentrate on the 359 firms that undertake product innovation activities.

It should be mentioned that unanswered questions among the surveyed firms along with firms which refused to answer, are likely to lead to biased sample selection. Firms that are willing and able to answer the questionnaires completely usually have a higher level of human capital or more formal organizational frameworks, which eases the understanding and communication between firms and the universities that conducted the survey. Moreover, these firms are more interested in the strategic development plan the research team promised to provide after the survey than the firms that refused or left too many questions unanswered, which reflects their upgrading-oriented strategy. In fact, this selection bias controls for the technological level of the surveyed firms to a certain degree, because it ensures that the survey firms' innovation activities are not limited to very low-value innovation, such as complete imitation without adaptation, and thus require more coordination and learning in the innovation process.

In light of proposition 1A, three aspects are taken into account throughout the interactive learning process: searching for information to facilitate innovation decision making (obtaining new product ideas), obtaining codified knowledge and obtaining tacit knowledge in the implementation process. Proposition 1C guides the design for questions on internal efforts when undertaking innovation activities. Furthermore, the scope of interactive learning is derived from proposition 1B and proposition 1D. Overall, Table 3.1 demonstrates the structural design of survey questions on innovation processes, including both internal efforts and interactive learning throughout the innovation process, as well as the ways that interactive learning is organized.

The intensity of the interaction with these players in the innovation process is measured by the firms' evaluation of the importance (from 1 to 5 with increasing importance). The firms are also asked to rank the importance of active searching, *Guanxi* with business partners as well as *Guanxi* with relatives and friends when they interact with these agents in the innovation process. Here, active searching refers to arm-length-market relation which based on pure contract relation, while

3.4 Empirical Results

Table 3.1 Structural indicators on innovation processes

			Remarks
Obtaining new product ideas	Internal efforts		*Own development of ideas*; Self Absorption and Learning through *license purchasing and reverse engineering*
	Sector product-field specific knowledge		Interacting with *parent companies, foreign customers, domestic customers*
	Generally applicable knowledge		Interacting with *universities, research institutes and sales agents*
Obtaining codified knowledge	Internal efforts		Self-purchasing of equipment and software
	Sector product-field specific knowledge		Interacting with *parent companies, foreign customers and domestic customers*
Obtaining tacit knowledge	Sector product-field specific knowledge	Active	Sending staff *to foreign customers or foreign lead firms, domestic customers or domestic lead firms*
		Passive	Receiving training and know-how from people sent *by parent company, foreign customers and domestic customers*
	Generally applicable knowledge		Sending staff *to universities* for training
Interaction mode	Informal *Guanxi* network		Interacting through *Guanxi*, for example gaining information on the reputation and capacity of innovation partners from other *business partners, relatives and friends* in the innovation process
	Active searching		Searching for information on partners *via Internet, exhibition and sales agents* in the innovation process

For detained information on question formulation, please refer to Sect. 7.1 Appendix A: Firm Questionnaire, Part C, question 27–30

Guanxi with business partners, relatives and friends refers to informal aspect of social relations. Although the interaction mode is not the main investigated aspect in this chapter, it will provide a first insight into the way that interactive learning is organized among the electronics firms in the PRD, China, which would pave the empirical evidence for further exploration on the way to secure and sustain interactive learning in the following chapters.

3.4 Empirical Results

3.4.1 Descriptive Results

At first, a general overview of firm characteristics concerning the innovation investment and the innovation outcomes among the surveyed firms is provided. The average age of the innovating firms was 8.8 years in the surveyed year 2009. Among the

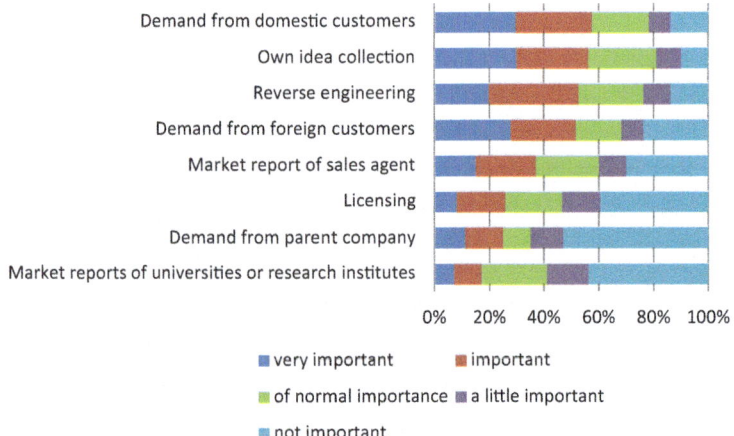

Fig. 3.1 Firm evaluation on ways of obtaining new product ideas

innovating firms, 8 % were large firms with no less than 300 million Yuan sale and no less than 2,000 employees. 39 % of firms had foreign participation. The median value of the share of innovation input in sales was 20 % in the first half of 2009, and one third of the firms invested over 20 % of sales in product innovation. Note that the survey year was during the recovering period of the financial crisis, so it indicates the stronger incentive to undertake innovation among electronics firms, even under market uncertainty. As for the innovation outcomes, more than 40 % of the firms surveyed had achieved significant or very significant improvement on product function expansion and product category upgrading.

From Fig. 3.1, it can be seen that electronics firms in the PRD rely very much on their own competence and reverse engineering to trigger innovation activities, indicating to some extent that firms in the PRD are increasing their internal absorptive capacity to transform externally codified knowledge, e.g. advanced product samples, into new product ideas and market opportunities. On the other side, the demands from foreign customers and domestic customers play a significant role in motivating the firms to undertake innovation. Compared to internal competence and closely linked partners in business operation, the impact of external knowledge-producing institutes, such as sales agents, universities and research institutes, on triggering innovation ideas is insignificant.

In the process of undertaking product innovation (Fig. 3.2), electronics firms in the PRD turn firstly to domestic customers for the support of equipment and software, secondly to foreign customers and finally to the parent companies, which corresponds to the aspect of triggering innovation ideas.

As shown by Fig. 3.3, the electronics firms in the PRD interact more with domestic customers to acquire technical experience and know-how, either in an active way (engineers sent to domestic lead firms or customers) or in a passive way (engineers sent by domestic customers). The interaction with foreign customers is the second most important channel of acquiring necessary tacit knowledge in order to undertake successful innovations. The other channels, such as universities and the parent company, have the least weighting in the interactive learning activities aimed at acquiring tacit knowledge.

3.4 Empirical Results

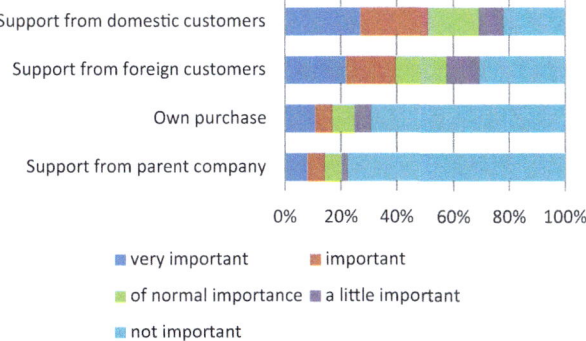

Fig. 3.2 Firm evaluation on ways of obtaining codified knowledge

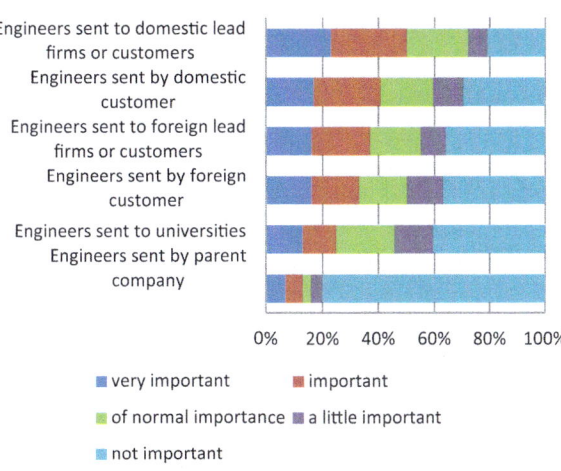

Fig. 3.3 Firm evaluation on ways of obtaining tacit knowledge

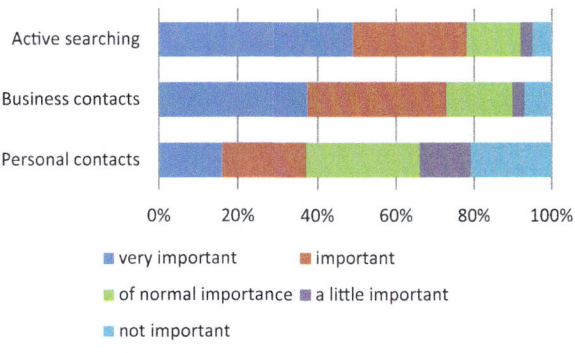

Fig. 3.4 Firm evaluation on interaction mode

In Fig. 3.4, it is shown that electronics firms in the PRD interact with external agents in innovation activities mostly through exhibitions, the Internet and sales agents, confirming an arms-length market relationship. Business contacts through recommendation by business partners are also widely applied. In contrast, the most informal personal networks such as recommendation by relatives and friends are rarely applied. In the process of interactive learning, informal relations are able to promote the level of information sharing and knowledge transfer due to trust between the partners. However, when the firm capabilities are not equally developed among the firms, interactive learning within informal frameworks tends to harm the incentives of firms with higher capabilities, and therefore reduces the frequency of interactive learning.

3.4.2 Econometrical Analysis

Due to the form of ordered responses designed in the questionnaire, latent class analysis is applied first to characterize clusters of similar cases in interactive learning behavior in product innovation processes. A latent class model groups the observations in terms of probability. This stands out from normal clustering methods, as it is able to provide measurements of parsimony and goodness of fit that are statistically sound. In this way, the subjectivity of decision about class number can be effectively controlled.

Table 3.2 shows the results of a latent class model where the 3-cluster solution is accepted. The fitness of this solution outperforms the 4-cluster solution owing to the more parsimonious and theoretically sound interpretability (for more information on the selecting procedures, see Sect. 7.2 Appendix B). The numbers in the table indicates the probabilities of firm's high evaluation on the importance of each aspect in the innovation process conditional to the respective cluster.

The three clusters identified differ in terms of the scope and intensity of interactive learning in the product innovation process. The first cluster is an underperformer in interactive learning activities. It groups together firms that are neither competent in internal learning activities, such as reverse engineering, licensing and triggering of innovation ideas by internal discussion, nor actively involved in interactive learning processes, such as getting innovation-related information and ideas and obtaining necessary codified and tacit knowledge for successful innovation outcomes.

The second cluster, which is referred to as the moderate interactive learning cluster, outperforms the weak interactive learning cluster in the aspect of applying interactive learning processes to trigger innovation ideas and obtain technical know-how and experience. Firms belonging to this group seem to interact more with domestic customers, or domestic lead firms in terms of obtaining tacit knowledge. Like the weak interactive learning cluster, the firms in the moderate interactive learning cluster mostly implement strategies of active searching and business networks when undertaking interactive learning. The probabilities of applying *Guanxi* networks for the second cluster are marginal (68 and 38% respectively with business partners and with relatives and friends).

The third cluster, which is referred to as the intensive interactive learning cluster, shows the greatest inclination to undertake interactive learning activities in the

3.4 Empirical Results

Table 3.2 Latent class model results for surveyed electronics firms

	Probability of high evaluation[a]	Weak interactive learning cluster	Moderate interactive learning cluster	Intensive interactive learning cluster
Obtaining new product ideas	Own idea collection	0.49	0.50	0.81
	Reverse engineering	0.45	0.48	0.77
	Licensing	0.14	0.24	0.56
	Demand from parent company	0.15	0.26	0.48
	Demand from foreign customers	0.38	0.51	0.83
	Demand from domestic customers	0.49	0.53	0.86
	Market reports of Sales agent	0.22	0.40	0.69
	Market report of universities or research institutes	0.03	0.17	0.49
Obtaining codified knowledge	Support from parent company	0.09	0.10	0.33
	Support from foreign customers	0.24	0.24	0.94
	Support from domestic customers	0.38	0.40	0.93
	Own purchase	0.19	0.06	0.27
Obtaining tacit knowledge	Engineers sent by parent company	0.08	0.09	0.26
	Engineers sent by foreign customers	0.16	0.29	0.76
	Engineers sent by domestic customers	0.24	0.41	0.80
	Engineers sent to foreign lead firms or customers	0.19	0.38	0.77
	Engineers sent to domestic lead firms or customers	0.35	0.53	0.78
	Engineers sent to universities	0.09	0.31	0.50
Interaction mode	Active searching	0.72	0.72	0.99
	Business contacts	0.65	0.68	0.97
	Personal contacts	0.25	0.38	0.59
Share of each cluster		50%	28%	22%

[a] Probabilities that the firm in each cluster give a high evaluation, i.e. important (4) or very important (5)

Table 3.3 Firm characteristics of the three clusters

		Cluster of latent analysis	Weak interactive learning cluster	Moderate interactive learning cluster	Intensive interactive learning cluster
Firm characteristic	Firm age (years until 2010)		9.84	12.22	11.37
	Firm ownership (% of foreign firms)		0.36	0.41	0.42
	Firm size (% of large firms)		0.08	0.05	0.12
	Export market (% of sales)		40.4	45.8	50.7
	Technical staff above bachelor degree (%)		33.75	32.61	42.15

innovation process, especially in the aspect of interacting with foreign and domestic customers to obtain innovation ideas and related codified and tacit knowledge. In addition, they also tend to interact with sales agents, universities and research institutes to trigger innovation ideas. Unlike the other two clusters, intensive interactive learners are inclined to apply *Guanxi* networks with business partners, relatives and friends, combining these with the active search strategy.

The class distribution indicates that interactive learning processes are still underdeveloped in the PRD, China. Half of the firms surveyed are still very weak in undertaking this kind of learning activity to take advantage of dynamic externalities, i.e. the knowledge spillovers from other firms. 28 % of the firms are nurturing the capability of interactive learning, while 22 % have shown the willingness and acquired the capability to undertake interactive learning to trigger product innovation and obtain support for necessary codified and tacit knowledge. The low frequency of the PRD's electronics firms in undertaking interactive learning reflects the immature internal absorptive capacity of most firms to understand and adapt knowledge from other agents effectively.

Table 3.3 shows the firm characteristics of each cluster in the aspects of firm age, size, ownership, market orientation and levels of human capital. The intensive interactive learning cluster has slightly more participation from foreign capital. Furthermore, there are more large firms in the group of intensive interactive learning cluster. The characteristics that makes the significant difference lie in the market orientation and levels of human capital. It turns out the intensive interactive learning firms are most export-oriented, which suggests the role of active integration into the global production networks in expanding the scope and intensity of firms' interactive learning activities. The share of technical staff having at least bachelor degree in intensive interactive learning firms also exceeds the other two clusters of firms by almost 10 %.

In terms of growth performance, as shown in Table 3.4, it is not surprising to find out intensive interactive learning firms underwent the least reduction in sales during the first half of 2009—the recovering period after the financial crisis, which again affirms the role of interactive learning in acquiring market information and

3.4 Empirical Results

Table 3.4 Growth performance of the three clusters

	Cluster of latent analysis	Weak interactive learning cluster	Moderate interactive learning cluster	Intensive interactive learning cluster
Growth performance	Sales growth (2007—first half of 2009)	−12.6	−5.6	−2.5
	Improvement on product function expansion (average score from 1 to 5)	3.6	3.6	4.1
	Improvement on product category upgrading (average score from 1 to 5)	3.4	3.8	3.9

reducing uncertainty. Moreover, they also lead other firms in terms of product innovation outcome, both for incremental functional expansion and discontinuous category upgrading.

The effect of interactive learning on the firm product innovation outcome is also tested. The independent variable for interactive learning is the binary response on whether the firm belongs to a specific cluster, and the benchmark is the weak interactive learning cluster. Control variables, including firm characteristics in terms of size, age and ownership as well as firm absorptive capacity in terms of CEO education, skill level of technical staff and initial product technology, are also integrated in the regression model, which are listed in Table 3.5.

The dependent variable in the regression takes the firms' evaluation on the degree of improvement (ranging from 0 to 5 with increasing significance of change)

Table 3.5 Control variables in ordered logit regression

Variable Name	Description
Firm size	Defined according to Chinese firm size standard, 1 as large firms with no less 300 million Yuan sales and no less than 2,000 employee, otherwise as small and medium-sized with the value of 0
Firm ownership	1 as firms with foreign participation (wholly owned or joint venture), 0 as firms with 100% domestic participation
Firm age	Years since establishment of the firm
CEO education	1 as CEO below bachelor degree; 2 as CEO with bachelor degree; 3 as CEO with graduate degree (master or doctor); 4 as CEO with bachelor or above combined with overseas experience
Skill level of technical staff	Percentage of technical staff that have bachelor degree or above *multiplied by* training frequency
Initial product technology	Defined according to International Standard Industrial Classification of all Economic Activities, Rev 3[a], 0 as producing low-tech products when starting business, 1 as producing medium- and high tech products when starting business

[a] Specific classification of products into the different levels could be referred to Sect. 7.3 Appendix C

in product function expansion and product category upgrading into account. It is considered that functional expansion and category upgrading are more complicated and thus require more interactive learning than improvement on design and quality. The product function expansion refers to the addition or upgrading of product functions within the same product category, while the product category upgrading refers to more discontinuous innovation, such as producing mainboard instead of network adapters, or producing mp3 players instead of flash memory. Due to the discrete and ordered feature of this multinomial-choice variable, the ordered logit model was applied. In order to reduce the dimension of the dependent variable that might influence the stability of the ordered logit model with a medium-sized sample, the dependent variable takes the average value of the evaluations on function expansion and category upgrading. In this way, the dependent variable measures the approximate evaluation of firms on the improvement of function expansion and category upgrading with increasing degree from 0 to 5.

Table 3.6 shows the result of the ordered logit regression.

Table 3.6 Ordered logit regression on product innovation outcome

Independent variables	Product Innovation outcome (Average score of evaluation)			
	Model 1 (Without control)		Model 2 (With control)	
	Odds ratio estimate	Coefficient estimate	Odds ratio estimate	Coefficient estimate
Moderate interactive learning cluster	1.243	0.217 (0.230)	1.073	0.070 (0.259)
Intensive interactive learning cluster	2.395	0.873*** (0.264)	2.710	0.997*** (0.298)
Firm age	–	–	1.020	0.019 (0.016)
Firm Size	–	–	0.739	−0.303 (0.439)
Firm ownership	–	–	0.572	−0.559** (0.222)
Skill level of technical staff	–	–	1.003	0.003** (0.002)
CEO education	–	–	1.355	0.304** (0.123)
Initial product technology	–	–	1.456	0.376 (0.242)
Prob > chi2	0.0038	0.0038	0.0000	0.0000
Pseudo R square	0.0122	0.0122	0.0518	0.0518
Number of observations	339	339	283	283
Proportional odds assumption test	$\chi^2 = 15.23, p = 0.02$		$\chi^2 = 36.17, p = 0.05$	

Standard errors in parentheses; $^*p < 0.10$, $^{**}p < 0.05$, $^{***}p < 0.01$

3.4 Empirical Results

The p-values of the chi-square likelihood ratio are all under 0.01, which guarantees that the model as a whole fits significantly better than an empty model. The coefficients in the ordered logit model display the ordered log-odds scale of change to higher order by one unit increase in the predictor while other variables are held constant. For example, if a firm were to apply intensive interactive learning in the product innovation process, the ordered log-odds of making better improvement on innovation outcomes would increase by 0.989 while other variables in the model are held constant.

Model one presents the results without control variables, in which the probability of having better improvement on product function expansion and product category upgrading for firms belonging to the intensive interactive learning clusters is more than twice as high, while for the moderate interactive learning cluster the positive impact is not significant.

Model two presents further results with the control variables, such as firm characteristics and firm absorptive capacity. Again, the intensive interactive learning firms possess a significantly higher probability of achieving better product innovation outcome than weak interactive learning firms, while the impact of belonging to moderate interactive learning cluster does not significantly improve the product innovation outcome compared to belonging to the weak interactive learning cluster. This verifies the hypothesis that the wide scope and high intensity of interactive learning activities contribute to innovation outcome due to the complexity and uncertainty of innovation processes.

The higher probability of achieving better innovation outcome when belonging to the intensive interactive learning clusters in model 2 compared to model 1 can be accounted for by the over-representation of foreign firms in the intensive interactive learning cluster, which has a significantly negative impact on the probability of achieving better product improvements. It is proved by model two that foreign participation would significantly reduce the probability of achieving better innovation outcomes. It can be concluded to some extent that foreign firms do not participate actively in product innovation activities, which is measured in the study according to function expansion and category upgrading. They might focus on the fields of R&D activities, patenting and scientific publication. However, if they undertake product innovation activities, they also rely on interactive learning to an even more intensified degree than domestic firms (See Table 3.3), in order to foster product innovation outcomes.

3.4.2.1 Robustness Test

For ordered logit regression, it is assumed that the relationship between each pair of outcome groups is the same. This is called the proportional odds assumption. In the last row of Table 3.6, the test of proportional odds assumption is demonstrated. For model 1, it has not sustained the test ($p=0.02$). Model 2 does not violate the assumption in a confidence level of 95%. However, it has violated the assumption if the confidence level is raised to 99%. The results indicate that the relationship

Table 3.7 Generalized logit regression on product innovation outcome

Independent variables	Probability estimates (1 to 2)	Probability estimates (2 to 3)	Probability estimates (3 to 4)	Probability estimates (4 to 5)
Moderate interactive learning cluster	19.690 (783.58)	0.998* (0.588)	0.053 (0.323)	−0.417 (0.363)
Intensive interactive learning cluster	0.946 (2.024)	1.543 (0.957)	0.871** (0.401)	0.867** (0.340)
Firm age	−0.083 (0.124)	0.009 (0.035)	0.059** (0.024)	0.012 (0.021)
Firm size	–	13.782 (1331.126)	−0.470 (0.592)	−0.212 (0.537)
Firm ownership	−4.985*** (1.591)	0.074 (0.504)	−0.381 (0.294)	−0.688** (0.307)
Level of technical staff	−0.015 (0.010)	0.004 (0.004)	0.003 (0.002)	0.003* (0.002)
CEO education	0.304 (0.673)	0.421** (0.266)	0.376** (0.158)	0.275* (0.156)
Initial product level	−4.004** (1.625)	0.486 (0.455)	0.439 (0.293)	0.503 (0.328)
Constant	8.626*** (2.744)	0.303 (0.712)	−1.199*** (0.459)	−2.269*** (0.512)
Prob > chi2	0.000			
Pseudo R square	0.128			
Number of observations	283			

Standard errors in parentheses; $^*p < 0.10$, $^{**}p < 0.05$, $^{***}p < 0.01$

between each pair of outcome groups does not hold for the same and one single model (model 1 and model 2) is not representative. Thereby, a generalized ordered logistic model is run to show the difference of the coefficients between each pair of outcome groups.

Table 3.7 shows the coefficients and significance level for each pair of outcome groups. Each model demonstrates the probability to move the evaluation to the next higher level (e.g. from 1 to 2, 2 to 3, and so on).

For the low-scale improvement on function expansion and category upgrading (i.e. from not significant to a little significant and from a little significant to normally significant improvement), the intensive interactive learning firms does not perform a significantly higher probability of achieving better innovation outcomes than the weak interactive learning firms. While for the high-scale improvement on function expansion and category upgrading (i.e. from normally significant to significant and from significant to very significant improvement), the widest scope and highest intensity of interactive learning displays a significantly important role in promoting the innovation outcomes. Meanwhile, the moderate interactive learning contributes to the innovation outcome when a firm is to scale up from making a little improvement to making normal improvement on production function expansion and product category upgrading.

It might be interesting to note that the domestic firms display significantly higher probability in fostering product innovation than foreign firms in low-scale product improvement ($\beta = -4.617^{***}$), but their superiority over foreign firms in promoting product innovation becomes insignificant (from 2 to 3 and from 3 to 4) or decreases a lot (from 4 to 5). This demonstrates the limited potential of domestic firms in promoting product innovation.

3.5 Discussion and Conclusion

Interactive learning is able to generate increasing return for the internal learning by doing and learning by using, creating positive externalities for the whole economy (Lundvall 2005). This chapter testifies to the complementary role of interactive learning to internal efforts both in assisting firms in acquiring information to make innovation-related decisions in an uncertain market, and supporting firms with necessary codified and tacit knowledge in problem-solving and knowledge exploration during the innovation process. Empirical investigation of electronics firms in one of the world's largest electronics clusters, the PRD region, China, highlights the importance of interactive learning with domestic customer in the production innovation process, while generally applicable knowledge is more attained through activities such as introduction of advanced samples and internal imitation process, leaving the channel of universities and research institutes rarely applied among the electronics firms.

In general, this chapter verifies that the scope and intensity of interactive learning contribute to better innovation outcomes. Based on the latent class model, it is possible to identify three clusters of firms that bear increasing degree of interactive learning activities. The third cluster, which covers one-fifth of the surveyed firms, undertakes the widest scope and highest intensity of interactive learning activities. Moreover, the intensive interactive learning firms have a much higher tendency to apply informal *Guanxi* networks, for example with business partners, relatives and friends, in the interactive learning process than the other two firm clusters.

The empirical results also demonstrate that both domestic firms and foreign firms embed informal social networks in interactive learning as a way to promote product innovation. However, the foreign electronics firms do not show great interest in undertaking product innovation in the PRD, China. Therefore, the Chinese government should encourage the foreign firms to be involved more actively in product innovation with measures such as tax reduction, subsidies and permits for domestic market access, because the interactive learning organized by foreign firms in the product innovation process does not only foster innovation outcome for themselves, but is also able to generate knowledge spillover from the foreign sector to the domestic sector.

The role of informal social assets in supporting interactive learning and fostering innovation merits deeper investigation. Informal *Guanxi* networks, which are widely applied in Chinese business modes, have been proved by many studies to have a positive role in reducing transaction costs and sustaining reliable and responsive supplier-customer relationships (Meyer et al. 2009; Wu and Choi 2004; Luo 2002; Zhou et al. 2003). However, the role of informal *Guanxi* networks in fostering interactive learning processes and boosting innovation outcome still remains unclear. Therefore, the consideration of informal *Guanxi* networks into interactive learning processes contributes further to the understanding of social factors that facilitates innovation in the Chinese context.

This chapter contributes to the understanding of the role of interactive learning in promoting innovation activities in the context of China, where innovation is presently viewed as the essential dynamics of economic growth in the face of external market change and domestic inflation pressure. Also, the empirical substances call upon research on the role of informal social networks in supporting interactive learning activities and its implication in the Chinese context. However, Boschma (2005) indicates that too much commitment to social networks might induce a lock-in effect and underestimation related to the risk of opportunism. In China, it can be expected that firms tend to resort less to social networks when stability-induced institutions are in place. Therefore, this study proposes a future research agenda to discuss the interactive learning pattern in a dynamic theoretical framework, calling upon particular attention to reveal the way it co-evolves with the maturing institutional framework conditions in transition China.

References

Arrow KJ (1962) The economic implications of learning by doing. Rev Econ Stud 29(80):155–173
Asheim BT, Coenen L (2005) Knowledge bases and regional innovation systems: comparing Nordic clusters. Res policy 34(8):1173–1190
Asheim BT, Gertler MS (2005) The geography of innovation: regional innovation systems. In: Fagerberg J, Mowery D, Nelson R (eds) The Oxford handbook of innovation. Oxford University, Oxford, pp 291–317
Asheim BT, Parrilli MD (2012) Interactive learning for innovation: a key driver within clusters and innovation systems. Palgrave Macmillan, Hampshire
Boschma R (2004) Competitiveness of regions from an evolutionary perspective. Reg Stud 38(9):1001–1014
Boschma R (2005) Proximity and innovation: a critical assessment. Reg Stud 39(1):61–74
Capello R (1999) Spatial transfer of knowledge in high technology milieux: learning versus collective learning processes. Reg Stud 33(4):353–365
Cooke P, Gomez Uranga M, Etxebarria G (1997) Regional innovation systems: institutional and organisational dimensions. Res policy 26(4–5):475–491
Cooke P, Gomez Uranga M, Etxebarria G (1998) Regional systems of innovation: an evolutionary perspective. Environ plan A 30:1563–1584
David PA, Foray D (1995) Accessing and expanding the science and technology knowledge base. STI Rev 16:13–68
Dosi G (1988) The nature of the innovation process. In: Dosi G, Freeman C, Nelson R et al (eds) Technical change and economic theory. Pinter, London, pp 221–238

References

Gjerde KAP, Slotnick SA, Sobel MJ (2002) New product innovation with multiple features and technology constraints. Manag Sci 48(10):1268–1284

Griffith R, Redding S, Van Reenen J (2003) R & D and absorptive capacity: theory and empirical evidence. Scand J Econ 105(1):99–118

Hage J, Alter C (1997) A typology of interorganizational relationships and networks. In: Hollingsworth JR, Boyer R (eds) Contemporary capitalism. The embeddedness of institutions. Cambridge University Press, Cambridge, pp 94–126

Howells JRL (1999) Regional systems of innovation? Cambridge University Press, Cambridge

Jensen MB, Johnson B, Lorenz E et al (2007) Forms of knowledge and modes of innovation. Res policy 36(5):680–693

Kline SJ, Rosenberg N (1986) An overview of innovation. In: Landau R, Rosenberg N (eds) The positive sum strategy: harnessing technology for economic growth. The National Academies Press, Washington, DC, pp 275–305

Krugman P (1991) History and industry location: the case of the manufacturing belt. Am Econ Rev 81(2):80–83

Levitt B, March JG (1988) Organizational learning. Annu Rev Sociol 14:319–340

Lundvall BA (1988) Innovation as an interactive process: from user-producer interaction to the national system of innovation. In: Dosi G, Freeman C, Nelson R et al (eds) Technical change and economic theory. Pinter, London, pp 349–369

Lundvall BA (1992) National innovation systems: towards a theory of innovation and interactive learning. Pinter, London

Lundvall BA (2005) Interactive learning, social capital and economic performance. Paper presented at the international conference on advancing knowledge and the knowledge economy, National Academies. 10–11 January 2005, Washington, DC

Lundvall BA, Johnson B (1994) The learning economy. Ind Innov 1(2):23–42

Lundvall BA, Johnson B, Andersen ES et al (2002) National systems of production, innovation and competence building. Res policy 31(2):213–231

Luo Y (2002) Partnering with foreign firms: how do Chinese managers view the governance and importance of contracts? Asia Pac J Manag 19(1):127–151

March JG (1991) Exploration and exploitation in organizational learning. Organ Sci 2(1):71–87

Marshall A (1920) Principles of economics. Macmillan, London

Meeus MTH, Oerlemans LAG, Hage J (2001) Patterns of interactive learning in a high-tech region. Organ Stud 22(1):145–172

Meyer S, Schiller D, Revilla Diez J (2009) The janus-faced economy: Hong Kong firms as intermediaries between global customers and local producers in the electronics industry. Tijdschr Voor Econ En Soc Geogr 100(2):224–235

Morrison A, Pietrobelli C, Rabellott R (2008) Global value chains and technological capabilities: A framework to study learning and innovation in developing countries. Oxf Dev Stud 36(1):39–58

Nelson CR, Siegel AF (1987) Parsimonious modeling of yield curves. J Bus 60(4):473–489

Nelson RR, Winter SG (1982) An evolutionary theory of economic change. Harvard University Press, Bonston.

Revilla Diez J (2000) The importance of public research institutes in innovative networks-Empirical results from the Metropolitan innovation systems Barcelona, Stockholm and Vienna. Eur Plan Stud 8(4):451–463

Romer PM (1986) Increasing returns and long-run growth. J Polit Econ 94(5):1002–1037

Salter WEG, Reddaway WB (1969) Productivity and technical change. Cambridge University Press, Cambridge

Simon HA (1957) Modles of man. Wiley, New York

Simon HA (1991) Bounded rationality and organizational learning. Organ Sci 2(1):125–134

Smith K (2000) Innovation as a systemic phenomenon: rethinking the role of policy. Enterp Innov Manag Stud 1(1):73–102

Storper M, Venables AJ (2004) Buzz: face-to-face contact and the urban economy. J Econ Geogr 4(4):351–370

Teece DJ (1986) Profiting from technological innovation: implications for integration, collaboration, licensing and public policy. Res Pol 15(6):285–305

Teece DJ, Pisano G, Shuen A (1997) Dynamic capabilities and strategic management. Strateg Manag J 18(7):509–533

Williamson OE (1981) The economics of organization: the transaction cost approach. Am J Sociol 87(3):548–577

Wu WP, Choi WL (2004) Transaction cost, social capital and firms' synergy creation in Chinese business networks: an integrative approach. Asia Pac J Manag 21(3):325–343

Yang C (2009) Strategic coupling of regional development in global production networks: redistribution of Taiwanese personal computer investment from the pearl river delta to the yangtze river delta, China. Reg Stud 43(3):385–407

Yeung HWC (2009) Regional development and the competitive dynamics of global production networks: an East Asian perspective. Reg Stud 43(3):325–351

Zhou X, Li Q, Zhao W et al (2003) Embeddedness and contractual relationships in China's transition economy. Am Sociol Rev 68(1):75–102

Chapter 4
Absorptive Capacity, Proximity and Innovation: Linking up the Intra-Firm Characteristic with Inter-Firm Linkages

Abstract Proximity concept provides a measurement of accessibility other than the concept of externality as just being there. To determine the impact of proximity on learning and innovation has become a key issue in economic geography. In the context of transition economy, this chapter focuses on the role of organizational proximity and social proximity in fostering complex product innovation activities with further consideration on sufficient absorptive capacity. Based on a set of questionnaires conducted in the PRD, China, this chapter investigates the capacity and strategy of local electronics firms to capitalize on proximity. The result shows that electronics firm, especially small and medium sized firms are more interested and capable to interact with domestic customers and external institutes to obtain tacit knowledge and trigger innovative ideas. Moreover, production experience in high-tech fields is an important component of absorptive capacity that enables the use of proximity, while the educated level of managerial staff and entrepreneurs that are able to negotiate and strategically couple with global firms is crucial in using organizational proximity to foster innovation. Finally, this chapter shows that a group of social active firms has emerged, which is essential for the formation of a dynamic regional innovation system. However, the effect of social proximity in fostering innovation is marginal. Finally, the chapter pointed out that governance infrastructure should be established to stabilize and support the interactive learning, especially that within the social proximity.

Part of the chapter has been reorganized and published as *Strategies of using Social Proximity and Organizational Proximity in Product Innovation: Empirical Insight from the Pearl River Delta, China. Fu W, Schiller D, Revilla Diez J. Zeitschrift fuer Wirtschaftsgeographie 56/1–2,* Copyright © 2012, Buchenverlag.

4.1 Introduction

In recent studies on economic growth, innovation and learning are considered as the primary dynamics, and territories instead of firms are becoming the foci of learning, particularly collective learning process among local agents (Lundvall and Johnson 1994). The literature of proximity provides a feasible analytical framework to understand the conditions for learning within and between different agents (Malmberg 1997; Boschma 2005; Menzel 2006). Due to the uncertainty of innovation process, innovation outcome strongly depends on firms' ability to capitalize on various proximities to facilitate learning and coordinate complex problem solving process.

Boschma (2005) classifies proximity into cognitive proximity, organizational proximity, social proximity, institutional proximity and geographical proximity. This chapter adopts this classification as the departure of analysis. In China, institutional framework such as formal rules and law is weak, and geographical proximity serves as a precondition for other proximities such as social proximity and institutional proximity to function. Therefore, this chapter would focus on the role of social proximity and organizational proximity in facilitating interactive learning and how cognitive proximity, which is measured as the firms' absorptive capacity in different dimensions in this chapter, influences the use of these two proximities.

On one hand, the literature of global production network suggests that organizational proximity is more accessible to firms organized under global flagship network, and this entitles them to access to particular product and process technology and intra-organizational know-how under the hierarchical control of the flagship (Ernst and Kim 2002; Gereffi et al. 2005; Yeung 2009; Whitford and Potter 2007). On the other hand, the literature of regional innovation system stresses the role of social proximity, which is often correlated with geographical proximity, in accelerating learning and constituting dynamic innovation synergies (Malmberg 1997; Porter 2000; Lazaarini et al. 2001; Asheim and Isaksen 2002; Iammarino and McCann 2006; Malmberg and Maskell 2006; Asheim et al. 2007). Furthermore, social proximity is widely applied as the informal inter-personal relationships in the research region the PRD, China (Zhou et al. 2003; Meyer et al. 2009) with regard to production activities. Consequently, the capacity to capitalize further on informal social relations in order to foster innovation is critical for the emergence of regional innovation system in the PRD, China.

In a dynamic global economy, local firms are required to reinforce social and organizational proximity mutually. On one hand, the capacity of local firms to capitalize on social proximity and to transform it into innovative synergy and profit gives higher incentive of foreign firms to transfer more advanced technology and activities to their organizational proximate partners in developing countries. Moreover, this entitles the local firms and governments more bargaining power to negotiate with foreign partners, which results in easier and more stable manipulation of strategic coupling. On the other hand, new technologies and market opportunities that are pumped into the local system by strategic coupling with distant partners strengthen the dynamic collective learning process in the region (Bathelt et al. 2004).

4.1 Introduction

Within the proximity approach, cognitive proximity is the basic one to gain other proximities. This implies that firms should develop certain level of absorptive capacity in order to capitalize on organizational proximity and social proximity. Cohen and Levinthal (1990) suggest that absorptive capacity is the primary requirement for firms to identify, interpret and exploit the new knowledge. This chapter identifies the absorptive capacity in Chinese firms embodied as the human capital, R&D activities and production experience in high-tech fields and further explores the influence of cognitive proximity, which is promoted through the strengthening of absorptive capacity, on the firms' capacity to use social proximity and organizational proximity in the interactive learning process.

Most of the firms in developing countries are weak in capitalizing on the proximities to foster innovation outcome due to the constraint of internal absorptive capacity. Usually, they rely on organizational proximity and social proximity only to gain market information and sustain reliable supplier-customer relationship. Therefore, this chapter aims to investigate the use of organizational proximity and social proximity in the process of product innovation, and compares their respective effect of fostering innovation outcomes. In addition, it tries to reveal their relationships with absorptive capacity of the firms. Moreover, distinction between large firms and small and medium sized firms in their willingness and ability to use different proximities would be identified as to provide more insight into the development of regional innovation system.

In sum, this chapter aims to address three questions:

1. Which components of absorptive capacity enable a firm to conduct extra-learning using either organizational proximity or social proximity in product innovation process?
2. What is the difference of small and medium-sized enterprises (SMEs) in terms of using proximity compared to large firms?
3. Controlling for other firm-specific characteristics, does the use of proximity contribute to product innovation outcome? And does the effect of social proximity differ from that of the organizational proximity?

In general, the empirical analysis of this chapter provides insight into the way of interactive learning in the product innovation process from the proximity perspective. The comparative investigation between the role of organizational proximity and social proximity in promoting innovation outcomes is able to offer insights into the development of regional innovation system which started from a mere FDI-driven platform in late 1970s.

The chapter proceeds as follows. Section 4.2 introduces the concept and taxonomy of proximity, and compares the role of social proximity and organizational proximity in supporting trust-based interactive learning, putting the theoretical discussion on a global-local scale. Section 4.3 analyzes the firm-specific factors influencing the absorptive capacity that is determinant in external interactive learning. For the theoretical discussion in these two sections, hypotheses are drawn. Section 4.4 presents the data and the applied methodology to operationalize the analysis and test the hypotheses. Section 4.5 discusses the results. To end up, Sect. 4.6 makes overall conclusions and draws future research direction.

4.2 Use of Proximity in Interactive Learning

4.2.1 Proximity: Concept and Taxonomy

The concept of proximity developed in 1990s by French School contributes to the understanding of the mechanisms that is working through the interactive process of knowledge transfer (Kirat and Lung 1999; Torre and Gilly 2000; Torre and Rallett 2005; Boschma 2005, Menzel 2006). Proximity is a concept that is usually discussed with innovation, since it plays an important role in promoting the trust and understanding when undertaking complex and highly risky innovation activities.

Proximity bears a plural sense. It goes beyond the geographical proximity, which has limited role without the support of other proximities. There are many classification of proximity in the literature. They are mainly associated with two perspectives: institutionalist approach and interactionist approach. The institutionalist approach concerns about three proximities: geographical proximity, organizational proximity and institutional proximity. In this approach, geographical proximity indicates the physical proximity without institutional context, while organizational proximity and institutional proximity bear institutional meaning in the way that the scope and scale of shared and common rules determines the function of proximity (Kirat and Lung 1999; Torre and Rallett 2005). However, interactionist approach only distinguish physical proximity (geographical proximity) from non-physical one, which is either defined by common resemblance or belonging to the same affiliation.

In some ways, the two approaches are overlapping, and it is possible to bridge over these two classifications (Carrincazeaux et al. 2008). In order to make the points clearer, this chapter adopts the classification developed by Boschma (2005), who tries to make a comprehensive understanding of various proximities. According to Boschma (2005), there are five different kinds of proximity considered:

- Cognitive Proximity: People have the same knowledge base and expertise can better learn from each other. For organizations which possess idiosyncratic nature of knowledge in the cumulative process of routine development, cognitive proximity rests on the similarities of technical and market competencies between the actors that affect the search and imitation cost when exploring new knowledge in other organizations.
- Geographical Proximity: It indicates the physical distance between the interacting actors measured by time or money, which depends on the infrastructure. Geographical proximity alone is not able to foster knowledge transfer and innovation. It might combine with cognitive proximity for firms to conduct in-time monitoring and comparing without direct interaction. It also strengthens the social proximity by offering more chances for face-to-face contact. As Howells (2002) puts it, the impact of geographical proximity is rather indirect and subtle.

4.2 Use of Proximity in Interactive Learning

- Organizational Proximity: It refers to the sharing of reference space and knowledge that is strengthened by hierarchy and control within the same organization (firm, group, cooperation network). With the development of information, communication and transportation technology, pure co-location is no longer determinant in knowledge transfer. Network, which even transcends the boundary of countries, begins to play a role as vehicles of knowledge diffusion.
- Social Proximity: It relates to trust and commitment based on kinship, friendship and cooperation experience. Social proximity does not only foster the communication of tacit knowledge which is difficult to be traded in the market, but also reduces opportunist behavior through the establishment of durable relations. It is often geography bounded because the geographical proximity enhances the chances of meeting and communicating.
- Institutional Proximity: Unlike social proximity which is based on informal social relations between agents at the micro level, institutional proximity is based on norms and values at the macro-level. It is depersonalized and relies on general trust, which is brought by common rules, norms and values that have been developed and established over a long term such as laws, regulations and cultural habits. Institutional proximity is also geography bounded at most of the time, but the scale might be larger than social proximity because institutions may exist in the level of town, city, province and country.

Proximity is not panacea. Too much proximity would lead to negative results such as too high factor price, lock-in effect, vicious competition and mistrust (Torre and Rallett 2005; Vicente and Suire 2007; Brossard and Vicente 2007). Boschma (2005) has summarizes the appropriateness of proximity, as summarized in Table 4.1.

Cognitive proximity serves as perquisite for learning, and it is easier achieved via the interaction within the supplier link and organizational boundary due to the continuity of knowledge transfer. Social proximity and institutional proximity rest often on geographical proximity to function properly. Capello (1999) demonstrates that social proximity and institutional proximity set in motion an informal and tacit transfer of information and know-how, which contributes to the transformation process for a specialized area to an innovative milieu. On the extra-local scale, organizational proximity gives a different meaning to supplier linkages, which facilitates the transfer of tacit knowledge by control and hierarchy.

In the context of China, which is undergoing a gradual transition towards a market economy, many formal institutions such as laws, regulations and organizations (work unions, research institutes, patent office, etc.) have already been established, although their enforcement is still problematic. Moreover, the institutional framework is unstable in the transition phase. As a result, the economic players in China do not tend to rely too much on institutions to do business (Zhou et al. 2003; Meyer et al. 2009). In the following discussion, I therefore do not focus on the role of institutional proximity at the macro level. Instead, the role of organizational and social proximity in fostering tacit knowledge transfer and dynamic innovative synergies is examined and compared.

Table 4.1 Appropriate distance of various proximities. (Boschma 2005)

	Too little	Too much	Solutions
Cognitive proximity	Less absorption	1) less to learn from	Diverse and complementary knowledge base of actors
		2) risk of lock-in	
		3) undesirable spillovers to competitors	
Organizational proximity	Strong tie (hierarchical): Control mechanism to ensure ownership rights and rewards when new knowledge creation is with uncertainty and opportunism	1) lock-in in specific exchange relations (esp. the relations are asymmetric)	Loosely coupled network
	Makes complex knowledge transfer more effective	2) lacks feedback mechanisms	
		3) lacks organization flexibility to implement innovation (interest group, vested interests)	
Social proximity	Lack of trust and commitment	underestimation of opportunism	Mixture of market and embeddedness
Institutional proximity	Weak formal institutions	Institutional lock-in and inertia	Institutional checks
Geographical proximity	No spatial externalities	Specialized region's case, but not geographical factor alone	Pipeline

4.2.2 Organizational Proximity and Social Proximity: Comparison and Dynamics

Figure 4.1 shows the how information and knowledge transfer across the firm boundary support the complex innovation process. The knowledge transfer organized within the social proximity and organizational proximity facilitates communication and strengthens cooperation owing to understanding and trust within the proximity boundary. Firms can on one hand interact and cooperate with organizationally proximate partners such as parent companies and foreign customers to gain information, ideas and supported knowledge, which surpasses the limit of geographical proximity. On the other hand, firms can also establish trust-based social network with the organizationally distant partners such as the domestic customers, universities, research institutes and market agencies, seeking the information

4.2 Use of Proximity in Interactive Learning 73

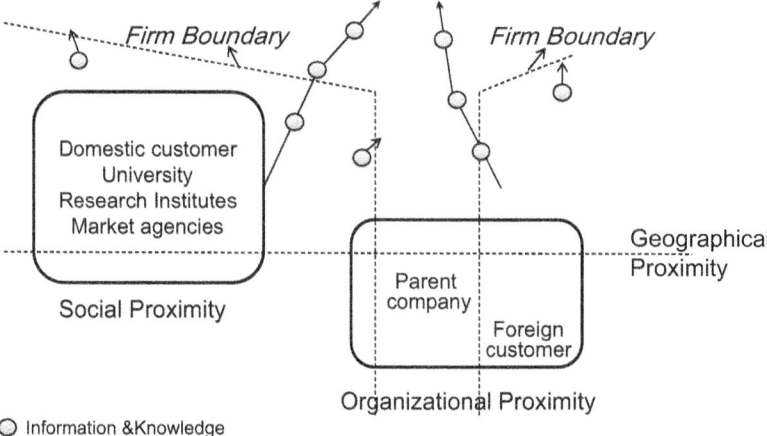

Fig. 4.1 Knowledge transfer within and across the firm boundary

and knowledge within the social proximity. Social proximity is usually geography bounded due to the positive role of face-to-face interaction in solving incentive problems, facilitating learning and providing psychological motivation (Storper and Venables 2004).

4.2.2.1 The Role of Organizational Proximity in Innovation in Latecomer Context

The literature on global production network has provided insight into the role of organizational proximity in industrial organization for multinational companies and its function as a vehicle of cross-country knowledge transfer. The global production network perspective pays attention to both intra-firm and inter-firm transactions and the respective forms of network coordination. With this network, information and resources flow rather easily among the flagship company, its own subsidiaries, affiliates and joint ventures, and the subcontractors, suppliers, service providers, and partners in strategic alliance (Ernst 2002).

The main purpose of this network is to provide the flagship, which includes the brand firms, contract manufacturer, first-tier supplier and large trade companies, with quick and low-cost access to resources, capabilities and knowledge that are complementary to its core competencies. Outsourcing of volume manufacturing enables these firms to combine cost reduction, product differentiation and quick response to the market (Ernst and Kim 2002).

Organization exists as a bundle of transactions or contracts (Coase 1937) and also as a bundle of knowledge (Barney 1991; Barney 2001). Following the logic of transaction cost minimizing, organizational boundary helps to curb the opportunist behavior of business partners, such as distorting business information, failing to fulfill commitments or malicious imitation, when rules and trust are absent.

Table 4.2 Governance of global value chain according to organizational proximity. (Gereffi et al. 2005)

	Organizational proximity: Close → Far				
	Hierarchy	Captive	Relational	Modular	Market
Description	Managerial control from headquarters to subsidiaries and affiliates	Small suppliers monitored and controlled by large firms	Relations managed through reputation and trust	Suppliers provide customer-specified products and "turn-key" services	Relations on contract specifications

Kindleberger (1964, pp. 146–147) argued that the industrial organization separating firms that deal with each other in arm-length market in the British case might be destructive in the way that impeded technological change, as part of the benefits from change cannot be fully internalized by each separated firm. Hennart (1993) suggests further that the organizational boundary remains when the internal managerial cost does not exceed the market transaction cost related to opportunism. On the other hand, the path-dependent nature of firm-routine development enables firm-specific tacit knowledge transfer more easily, which reduces internal management cost to a large extent.

Thereby, the governance of global value chain depends on three factors: (1) the complexity of transactions; (2) the ability to codify knowledge; (3) the capabilities in the supply-base (Gereffi et al. 2005). In another word, the principle of exerting organizational control on distant subsidiaries and suppliers is that the more complex the transactions, the harder it is to codify knowledge and the lower capabilities the supply-base has, the more organizational proximity is needed. These three aspects are closely related to the opportunism risk and transferability of knowledge.

According to these three standards, Gereffi et al. (2005) define five types of global value chain governance: hierarchy, captive, relational, modular and market, which range from high to low levels of explicit coordination and power asymmetry. In Table 4.2, I summarize the characteristics of these five governance forms based on Gereffi et al. (2005).

For the global flagship that organizes production in developing countries, organizational proximity is conductive to reducing opportunist risk related to physical and human capital investment. Global flagship takes on the responsibility of investing in setting up and upgrading machines as well as training skill in the beginning of operation in developing countries due to the underdevelopment of technology and human skills in these regions. In addition, hierarchy and control within the same organization enables the efficient downward transfer of knowledge. Many studies on developing countries have shown that most of the firms conduct innovation "in-house" instead licensing and assigning contractual arrangements to unaffiliated firms (Schmitz 1995; Schmitz and Nadvi 1999). One of the explanations offered by Teece (1986) are that proprietary considerations are assisted by organizational integration, since contracts, proprietary rights and technology transfer via the market are complex and, especially in developing countries, often too expensive.

4.2 Use of Proximity in Interactive Learning

In short, when the institutional environment, consisting of elements such as property rights and related business laws, is not fully developed, and the embeddedness of global firms is not mature enough to ensure social proximity that brings mutual trust, flagship companies tend to restrict knowledge to flow only within the firm boundaries in order to reduce the chances of opportunism and ensure return on internal R&D efforts.

Accordingly, for firms in latecomer countries, organizational proximity provides two advantages:

Firstly, it provides them the access to knowledge, especially tacit knowledge in advanced technological field. The flagship typically provide subsidiaries or closely cooperating suppliers with encoded knowledge, such as machinery, blueprints, production and quality control manuals, product and service specification and training handouts. For latecomer firms, they have much more profit room than lead firms by just sourcing the existing technology to push the internal technological frontier. Yeung (2009) state the importance of external network building in acquiring capabilities in the Asian context, and introduces a concept called "strategic coupling" to better understand the evolution of local and regional firms in their dynamic articulation in global production network. Morrison et al. (2008) also found that firms gain technological capabilities from participating in global value chains. Not only that, latecomer firms also join the international production network in order to acquire tacit knowledge to absorb the encoded knowledge by having the engineers and managers from foreign partners train on site. However, Ivarsson and Alvstam (2005) demonstrate that geographical proximity to the foreign transnational corporations is crucial for local suppliers to absorb external technology through regular and ongoing interaction with their primary foreign customers.

Secondly, reliance on organizational proximity prohibits involuntary knowledge spillover between organizationally distant firms when the cognitive proximity between clustering firms is too little. Because of the abundance of labor force in China, it undermines the incentive of firms to stick to long-term technological and managerial upgrading activities in the face of large-volume production demand from developed countries. Low-cost and flexible responding strategy becomes the general choice. Many firms in the specialized areas in China compete fiercely in low-tech product field with low price and flexibility advantage. Because of the standardization of most of these low-tech products, the idiosyncratic nature of knowledge is rather insignificant and little complementary knowledge can be shared between them, which all lead to a too close cognitive proximity. In this case, firms are reluctant to share knowledge because the imitation cost is rather low. By contrast, firms stick to organizational proximity to source external knowledge to support complex innovation and upgrading activities.

In sum, global flagship companies tend to use organizational proximity to reduce opportunism when institutional proximity and social proximity cannot be assured, and local firms tend to rely on organizational proximity to gain access to knowledge and prevent involuntary knowledge spillover when firm-specific routines has not yet fully developed and diversified to ensure an appropriate cognitive distance between local firms.

It is possible that the suppliers upgrade and co-evolve with the buyer when the technological and organizational change enables a more sophisticated supply chain (Yeung 2009). In 2004, Lenovo bought the PC operation from IBM and upgraded from an OEM to an original brand manufacturing (OBM) producer. In 2004, TCL (Shenzhen) co-established a mobile phone joint venture with Alcatel. In 2007, China Electronic Cooperation subsidiary Sungfei (Shenzhen) acquired the mobile phone operation from Phillips. These are all examples of upgrading by enhancing internal absorptive capacity and strategically recognizing the coupling chances with leading global firms.

However, organizational proximity alone has a limited role in upgrading and innovation. Firstly, many brand owners arrange the global strategic layout in such a way that strategic R&D, marketing and management are located in their home countries or in regions in developed countries where innovation partners and reliable institutions are available (Feinberg and Gupta 2004), while functions such as production, sales and logistics are located in developing countries (Pan and Chi 1999). Although the internationalization of R&D activities has grown significantly since 1990s (OECD 1998), technology and knowledge to which domestic firms have access is still limited and mostly low-end. Secondly, global buyers tend to promote incremental product and process upgrading and oppose upgrading if this creates opportunities for suppliers to acquire a broader range of customers (Humphrey 2004). Consequently, the global buyers and traders might be by-passed by suppliers if the latter gain the ability to work directly with brand companies in developed countries.

In the electronics industry, there is a trend of applying less hierarchy relations in global chain governance mode. Maturing technology such as module production that enables the codifiability of knowledge is one of the reasons behind this trend. The rise of contract manufacturers in the PRD, China displays a massive shift towards large-scale vertical re-integration that offers the one-stop buying services for many brand companies (Luthje 2004). In a sense, the growing capability of local firms shortens the cognitive distance with global flagships and, at the same time, widens the cognitive distance between local firms. For local firms in a gradually maturing industrial cluster, they possess the benefit of use social proximity due to co-location and cultural similarity. Thereby, I would like to turn to the role of social proximity in fostering innovation as a means of overcoming the shortcomings of organizational proximity.

4.2.2.2 The Role of Social Proximity in Innovation in Latecomer Context

Social ties and relations have an influence on economic outcomes (Granovetter 1985). Social proximity is secured through informal daily face-to-face interaction such as meeting, chatting, and eating together and joint entertainment. Trust and commitment are gradually established in the social interaction process, which contributes to interactive learning and cooperation. Social networks are not spatially bounded, but it can be sustained and produced by the ongoing collective interaction of player located close to each other (Boschma 2005). It is worth mentioning that social proximity differs from institutional proximity. In the case of the former, people build trust in each other due to continual interaction and a deeper understanding in daily life, rather than in the latter, where common sets of values and recognition of rules are the key factors (North 1990; Boschma 2005).

As discussed before, many firms in specialized clusters of developing countries compete fiercely in low-end production. Because of the standardization of most low-tech products, firms are reluctant to share knowledge as the reciprocity of interaction is low and the risk of imitation is high. In this case, social proximity, such as that between customers and suppliers, is only used as a way of sustaining an agile and responsive production system. As a result, the role of social proximity in fostering innovation is limited, which leads to rather loose local innovation networks.

The socially and territorially embedded collectively interactive learning process is becoming prominent feature of competitive industrial clusters even in a globalized era (Maskell 1998; Asheim and Isaksen 2002). The approach of regional innovation system takes the regionalized assets and processes as the primary account for the innovation capabilities of the firms (Cooke et al. 1997; Doloreux and Parto 2005). In a well-functioning regional innovation system, the local firms are capable of capitalizing on social proximity not only to facilitate effective knowledge transfer, but also to generate innovation outcomes.

Guanxi, as an informal way of doing business in China, has received growing attention in the recent organizational literature (Park and Luo 2001; Ramasamy et al. 2006; Zhang and Zhang 2006). Similar to the concept of social proximity, *Guanxi* refers to the informal interpersonal relationships and exchange of favors for the purpose of doing business in traditional Chinese society (Lovett et al. 1999). There are three major categories of *Guanxi*: obligation and loyalty to family members or relatives—defined as the obligatory type of *Guanxi*, mutual assurance to friends, classmates and colleagues—defined as the reciprocal type of *Guanxi*, and understandings with acquaintances—defined as the utilitarian type of *Guanxi* (Zhang and Zhang 2006). Peng et al. (2008) points out further that the reciprocal and utilitarian types of *Guanxi* are becoming more important than the obligatory type in later phases of the institutional transition. In reciprocal *Guanxi* between friends and colleagues in particular, the implicit rule of "paying back favors" (Chinese term: *Renqing*), due to the fear of damaging one's social reputation and prestige, actually strengthens the constant social interaction through the idea of exchanging favors. Generally speaking, *Guanxi* in China is a common practice and is even more complicated than any kind of Western interpersonal relationship, since the Chinese have been more or less unintentionally or unconsciously involved in complex *Guanxi* networks ever since they began their working and social lives.

From the organizational perspective, local firms tend to apply an informal network-based strategy in the uncertain environment in China (Peng et al. 2008). Due to the gradual approach in the transition, many institutional setups have been subverted and not yet substituted, which has resulted in institutional loopholes. As a result, the legal system, property rights protection, industrial regulations and standards are all underdeveloped. Furthermore, the transparency and corruption issues have created an unreliable institutional environment. Under these circumstances, people tend to resort to *Guanxi* whenever problems emerge.

In the Chinese business world today, *Guanxi* plays an important role in facilitating economic exchanges and overcoming administrative costs in the face of a deficient institutional framework (Park and Luo 2001), when starting the business, concluding contracts, acquiring institutional protection, and responding flexibly to changing demands. However, its role in innovation has not yet been analyzed. In fact, there are several aspects in which *Guanxi* can function to promote communication and innovation synergies among firms with sufficient internal capabilities.

The precondition of interactive learning for the purpose of innovation is the sharing of information, knowledge and ideas, as well as commitment and loyalty in the investment phase. In this case, there are risks of opportunism, i.e. asset specificity, and behavioral and environmental uncertainty (Standifird and Marshall 2000). In summary, *Guanxi* curbs the risk of opportunism related to innovation in the following ways.

First of all, *Guanxi* with business partners can reduce the risk of asset specificity, which refers to the circumstance in which partners, who do not own and invest specific assets, switch suddenly to other partners in the process of innovation. The essence of *Guanxi* lies in the Confucian thought of harmony and an orderly world. Reciprocal *Guanxi* with business partners is path-dependent to some extent, because people are less disposed to ruining the precious *Guanxi* networks for quick profit. Long-term *Guanxi* acts as a constraint for opportunism, and this brings mutual trust and assurance in the cooperation process.

Secondly, *Guanxi* networks with other partners can reduce the risk of behavioral uncertainty when sharing knowledge and ideas with cooperation partners. As an old Chinese saying goes, "you will never be defeated if you know everything about your opponent". For example, if the cooperation partner has the potential to steal your ideas in order to develop a new product ahead of you, and the contracts and legal systems are not able to help or cost too much, it is safer to know ex ante about the background, reputation and capacity of your cooperation partner through the *Guanxi* network with other managers (as intermediaries) in the industry. Meanwhile, *Guanxi* network with other managers, who know well about the cooperating partners, can be also used as a way to curb the opportunist behavior through reputation and popularity in the complex web of relationship for specific social circle.

Thirdly, *Guanxi* with government officials can reduce the risk of environmental uncertainty, as innovation policies are always unsteady and vague in China. Mangers and entrepreneurs cannot simply rely on government bulletins as their information channel. They actually rely more on *Guanxi* for information searches and authentication. They often obtain key information and detailed explanation of the policies through obligatory or reciprocal *Guanxi*. Information transferred within the *Guanxi* networks is more reliable and trustworthy, thus facilitating the managers' decision-making on investing in innovation projects. Moreover, *Guanxi*, i.e. being related to or befriended with government officials provides access to scarce resources such as innovation funds and high-end technology transfers, because government officials in China exercise personal preference in the selection process in lieu of strict regulations and market mechanisms

However, *Guanxi* networks carry the risk of a negative lock-in effect. As *Guanxi* networks depend on the constant exchange of favors, it is also fragile once the exchange stops. Firms are locked in with current business partners, fearing a destruction of the subtle *Guanxi* network with a single business partner and all other partners who are related to this partner. In this case, firms do not act as profit-maximizing entities, but rather as *Guanxi*-satisfying ones. Outdated production modes and product types might persist and are harmful for upgrading and innovation (Saxenian and Hsu 2001).

One of the disadvantages of *Guanxi* is that it might damage the development of firm internal capability due to the limit of time and resources. While the Chinese enjoy the benefit of *Guanxi*, they also bear the *Renqing* burden, that is to say, they also take the reciprocal obligation and must repay it in the future (Luo 1997). Therefore, the *Guanxi* network costs time and money. It is a complex weaving interpersonal net that requires constant monitor, investment and subtle utilization. As time and resources are limited, gain in *Guanxi* network improvement must lead to the lack of investment in other aspects such as managerial capability and technological capability. Su and Littlefield (2001) distinguish *Guanxi* into two forms: favor-seeking *Guanxi* and rent-seeking *Guanxi*. As discussed before, favor-seeking *Guanxi* strengthens the continuing social interaction between firms and would contribute to knowledge transfer and innovation when the firms develop enough internal capacity to capitalize on it. However, rent-seeking *Guanxi* harms the overall efficiency of economies. Resource is distributed in accordance to *Guanxi* with government officials instead of capability and efficiency of the firms. This actually suppresses the firms' incentives to invest in managerial and technological upgrading, and leads to overinvestment in *Guanxi* network. As a result, Pareto efficiency of the whole economy would be reduced.

As a result, the importance of internal absorptive capacity should be highlighted as the support for using social proximity in a sustainable way. Only when the firms develop enough capacity to absorb, adapt and exploit the information and new knowledge, can social proximity (*Guanxi*) contribute to the innovation and growth of the firms in the long term.

4.2.3 Proximity for the SMEs in the Clusters

There is an on-going discussion on whether spatial fragmentation process and the development of global production network would undermine the localized external economies (Piore and Sable 1984; Storper 1995; Coe et al. 2004). The experience of "Third Italy" draw the scholars' attention again to the localized economies sustained among the small and medium sized firms (Piore and Sable 1984). In Third Italy, flexible and intertwined agglomeration of small and medium sized firms is able to stimulate the learning required for product and process innovation (Whitford and

Potter 2007). Capello (1999) demonstrates that collective learning in the local scale is the main way of achieving new resources for SMEs. Owing to SME's prominent role in a clustering, it is important to discuss how SMEs undertake interactive learning because it is of great relevance to the formation of a regional innovation system in which reciprocal innovative synergies among SMEs feed the system with growth dynamism and resistance to violent market change owing to the cushion effect.

The factor influencing the ability and willingness of different sized firms to use proximity is twofold: (1) resource and capability restriction of small firms; (2) the lack of interest of large firms in exploiting minor profit.

- Resource and Capability restriction

  Innovation activities require financial commitment on behalf of the firms to build up competences and skills through training, engineering activities and information search (Goedhuys 2007). Product innovation even requires non-deployable equipment investment which leads to asset specificity. The internal cash flow comes from the degree of market power a firm possesses, so that degree of vertical integration provides ample availability of capital (Armour and Teece 1980). Moreover, the significant volume of production capacity of large firms enhances the negotiation power with customers, and it renders them not to trap in one single customer. In China, OEM producers have to accept the "account receivable" capital chain mode owing to fierce competition based on flexibility (Smith and Schnucker 1994), implying that availability of multiple customers is able to contribute to increasing internal cash flow. Therefore, large firms on the contrary own more resources to conduct internal knowledge creation such as purchasing specialized machines, skill training and R&D activities and are less willing to grasp advantage of socialized knowledge transfer and creation as they possess the management capability to control opportunism in the innovation process by internal activities.

  On the contrary, small and medium sized firms are usually young and have not yet developed mature firm-specific routines and capabilities, especially in managing business and people. Due to the immature management capability, domestic firms are not able to internalize many functions and transactions within the firms to avoid external uncertainty as the large firms do. Furthermore, the constraint of internal cash flow makes the innovation cooperation such as sharing machines and key skills for SMEs indispensable. In this case, social proximity, embodied as *Guanxi* in China, displays as a favorable substitute for organizational proximity to build trust and control opportunism in the cooperation.

- Lack of interest of large firms in exploiting minor profit

  Scherer (1998) points out that an overlooked strength of small firms in exploiting innovation as the "excitement" of exploiting something new, which results from the sophistication of technological advance leading to myriad of narrow and detailed innovation such as on fabrication, material and minor components. Corporations with giant profit, on the contrary, feel less appeal in exploiting the profit of making too modest changes than small firms.

Nevertheless, the modest change in one part induces systematic change in the whole product series. By using social proximity, small and medium sized firms are able to work through technical problems and respond to the market change expeditiously through the interaction with users, service providers and other knowledge-intensive organizations.

However, homogenous product lines or markets might harm the reciprocal principle of cross-organizational knowledge transfer and dynamic innovative synergy among SMEs. Once the socialized process of knowledge creation and innovation cannot be guaranteed, the survival and growth of small and medium sized firms is in danger. In this case, according to Capello (1999), regional production system remains in the phase of industrial district, where social proximity only strengthens trust in supplier-customer relationship and reduces transaction cost among them. The higher level of regional production system as an innovative milieu cannot be achieved without the function of social proximity in promoting dynamic innovative synergies among local firms, especially the small and medium sized ones. Moreover, it should be noted that the advantage of small enterprise spatial systems tend to exhaust or even vanish in the face of dramatic technological change and macro-scale structural transformation, as evidenced in the difficulties of coping with post-industrial transition for the Third Italy beginning in the late 1990s and lasting all the way to 2000s (Bianchi 1998; Hadjimichalis 2006).

4.2.4 Brief Summary

For firms in latecomer countries, organizational proximity is of particular importance. In the early phase of development, the capability of local firms is not fully developed due to the weak industrial base, thus resulting in an ill functioning knowledge spillover mechanism in the local scale. In this phase, control and governance in the same organizational framework by foreign parent company or OEM customer is essential in organizing production in the region, and this becomes the main source for local firms to get codified and tacit knowledge, mostly in a passive way. However, organizational proximity in this phase is not able to trigger innovation with the absence of appropriate cognitive proximity between the foreign firms and local firms.

With the development of local production system, local firms have accumulated a certain level of capability which finally enables them to absorb and exploit new knowledge. In this circumstance, firms can either use organizational proximity to seize the profit opportunities of value chain upgrading with the sophistication of supply chain and technological diversification, or can they capitalize on social proximity to form reciprocal innovative synergy with organizationally distant partners that possess the appropriate cognitive distance. Particularly for small and medium sized firms, the collective learning facilitated by social proximity is essential for their survival and growth, and is also important for the development of a self-sustained local production system (Capello 1999).

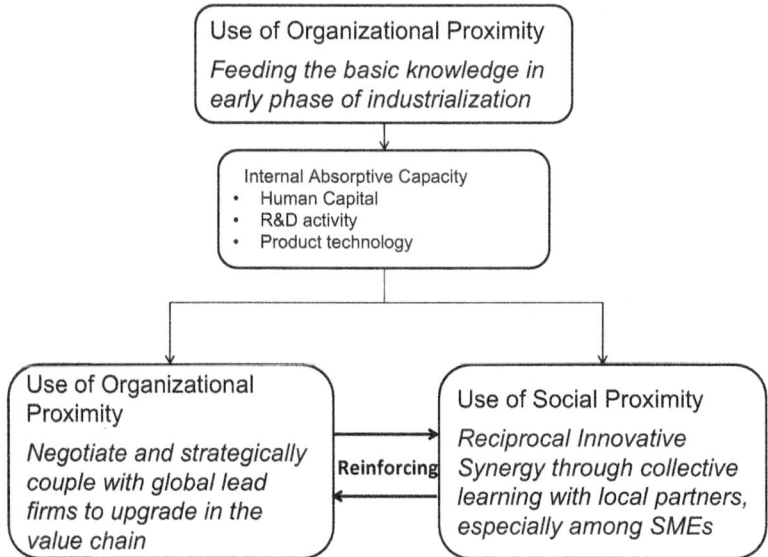

Fig. 4.2 Casual interaction of organizational proximity and social proximity in regional development

The use of organizational proximity and social proximity by firms are mutually reinforcing. On one hand, ability of local firms to use social proximity and transform it into innovative synergy and profit gives higher incentive for foreign firms to transfer more advanced technology and activities to their organizational proximate partners in developing countries. Moreover, this entitles the local firms and governments more bargaining power to negotiate with foreign partners, which results in easier and more stable manipulation of strategic coupling. On the other hand, new information on market and technology that pumps into the local system by firms using organizational proximity with geographically distant partners renders the local collective learning more dynamic (Bathelt et al. 2004).

Figure 4.2 illustrates this causal interaction concerning with the use of organizational proximity and social proximity in different phases of regional development.

Based on the discussion on the role of organizational proximity and social proximity on capability development and innovation for the firms and regions, I draw the following hypotheses:

Hypothesis 1 By developing absorptive capacity and strategic coupling within the global production network, it is possible for latecomer firms in emerging regions to capitalize on organizational proximity in order to foster innovation and upgrading. However, firms that rely only on a vertical hierarchy with global lead firms to foster innovation have limited potential for upgrading their position in the value chain.

Hypothesis 2 Most Chinese firms are engaged in *Guanxi* networks, which is an ongoing mode of interaction for maintaining social proximity between business partners. Firms with limited capabilities and short-term strategies are only able to capitalize on *Guanxi* for low-cost and flexible production. On the other hand, in a mature regional innovation system, firms are capable of using social proximity to facilitate the complex interaction in the innovation process and to upgrade their position in the value chain.

Hypothesis 3 Small and medium sized firms are more inclined to use social proximity as a strategy for knowledge creation and innovation than large firms, and their capacity in using social proximity is the primary element of a well-functioning regional innovation system.

4.3 Absorptive Capacity in the Firm Level as Precondition of Interactive Learning

Internal absorptive capacity and external interaction are complementary elements for successful learning (Cohen and Levinthal 1990; Arora and Gambardella 1990; Gambardella 1992; Tripsas 1997). Cohen and Levinthal's (1990) seminal work on absorptive capacity points out that certain capability should be developed to profit from external knowledge. Moreover, it does not only contribute to the successful absorption of external knowledge, but also promote the investment in exploring new domains of knowledge and new market. Cohen and Levinthal (1990, p. 137) implicitly elucidate the latter respect by saying that higher level of absorptive capacity enables the firms to be more sensitive and proactive to opportunities present in the technical and market environment.

Zahra and George (2002, p. 185) conceptualize absorptive capacity as "dynamic capability", which pertains to knowledge creation and utilization that enhances the firm's competitive advantage, and it composes a process of acquisition, assimilation, transformation and exploitation. Absorptive capacity can be viewed as the cognitive structure of firms that support the interaction with external partners, and Cohen and Levinthal (1990) identify this cognitive structure as the prior related knowledge of firms.

In this sense, absorptive capacity is closely related to the firm routine (Nelson and Winter 1982), which refers to particular experience and know-how that the firm accumulates over time. Current routines does not only influence the strategies the firms adopt, the activities they take and the opportunities they perceive (Boschma 2004), but also, in another way around, influence the efficiency of processing information and learning. As emphasized by Cohen and Levinthal (1990) in the discussion of absorptive capacity, it is rather an intangible concept and can be only indirectly measured. Nieto and Quevedo (2005) have comprehensively reviewed the empirical works on absorptive capacity, in which the measurements of absorptive capacity mainly include R&D activities and its linkage to basic research, patents, technical staff, product characteristics and management practice. In this section, I would introduce variables that are relevant and comparable among the firms in the context of latecomer firms. These variables constitute the cognitive structure of firms in perceiving innovation investment opportunities and fostering innovation outcomes. They are human capital, R&D investment and production experience in high-tech fields.

4.3.1 Human Capital

In macroeconomics, many scholars argue that skill should be accumulated at a high rate when technology is imported from outside in order to support sustained growth (Michaely et al. 1991; Lall 1992; Keller 1996). However, human capital accumulation has been proved not to directly determine the economic growth rate (Benhabib and Spiegel 1994). Rather, it influences the economic growth by attracting physical capital accumulation and affecting the speed of absorption of technology (Nelson et al. 1966; Lucas 1990).

Individual talent is the basic element in an organization. Beyond the individual capacities, what is more important for an organization is the sum of the individual capability, or, put it in another way, a collectively organizational learning process that aims to rightly allocate individual capabilities to tasks and maximize them. In this respect, organizational routine and culture plays a key role in combining, managing and driving the mobilization of individual capability (Shein 2004).

Departing from this argument, I relate human capital to the capabilities of technical staff, managerial staff and entrepreneurs. For technical staff, they directly involve in production and technological innovation and their individual capabilities make up the core element of the cognitive structure of the firm. On the other hand, the managerial staff and entrepreneurs are responsible for the optimal allocation of individuals to tasks and also shoulder the role of gate-keeper for the organization.

Technical staff is the main actor to exchange the know-how with external partners to a large extent, because their knowledge in specific domain enables them to recognize and value new related knowledge (Carter 1989). Technical staff does not only play a role in acquisition process, but is also determinant in assimilation, transformation and exploitation process, because they are the main carrier of tacit knowledge through years of team work. By team work, I mean that it is not only the individual capabilities of the technical staff that matters for the learning process, but also the common recognition of organizational codes and technological routines.

The collective learning and knowledge creation process between the technical staff also relies on the managerial capability to allocate the individual capabilities. Human resource management has been applied to define this capability and it has been also proved to exert a positive impact on innovative performance (Michie and Sheehan 1999; Laursen and Foss 2003). Vinding (2006) goes further to support the positive relationship between human resource management and radical innovation. Managerial staff's ability in fostering the organizational culture, optimizing the organizational structure and motivating the qualified staff is thus essential in regard to the assimilation, transformation and exploitation of external information and new knowledge. Moreover, they are also "boundary-spanning" actors between the departments and facilitate the transformation process of external knowledge within the firm (Allen 1977; Cohen and Levinthal 1990).

The human capital in the firm level does not only rely on the quality of the public educational system, but also on the internal development process. The public educational system mainly instructs generic knowledge that paves the way for the career development of graduates, and it is actually the firms' internal training efforts that organize the staff to collectively develop firm-specific competency and to achieve competitive advantage (Becker 1964; Barney 1991). Cohen and Levinthal (1990, p. 135) assert that the internal staff should be not only competent, but also quite "familiar with the firms' idiosyncratic needs, organizational routines, and extramural relationships", so that they are able to integrate complex external knowledge into the firms' activities. Thereby, the training effort in the firm plays a more important role in implanting the codes and routines in the technical staff that facilitates the communication and exploiting process within the organization than enhancing the individual capabilities of technical staff. However, if the gain of productivity and performance does not exceed the training investment as well as the monitoring cost of "poached talents" (Williamson 1975; Tsang et al. 1991), the firms tend to invest less in internal development efforts and turn to the market.

Besides technical and managerial staff, Schumpeter's (1942) seminal work on innovation stresses the role of entrepreneur as the agent of "creative destruction", implying that they are active in introducing new product, new technology and new combination. Seeing from the perspective of a firm, the founder or CEO often assumes the tasks of initiating new product development by a holistic thinking of firm-specific capability, market trend and network availability. Higher educational level of entrepreneurs entitles them more capability to negotiate with external actors. Especially for latecomer firms, overseas educational and working background of entrepreneurs can bring about more opportunities of value chain upgrading by grasp of market trends and the language and cultural skill in negotiating with global partners. Saxenian and Hsu (2001) found that US-educated Taiwanese has coordinated the process of reciprocal industrial upgrading between Silicon Valley and Hsinchu Park in Taiwan. Moreover, the CEOs often keep intense interaction with the companies and the business partners that they served before. Network relationships that the entrepreneurs have established in the past working experiences constitute valuable assets for their current entrepreneurial activities. Romijn and Albaladejo (2002) find a positive relationship between the founder's work experience in either multinational or large domestic firms and the firm's innovative capabilities in the UK.

4.3.2 R&D Activities

Owing to continuity of knowledge transfer and tacitness of knowledge, knowledge can be only easily acquired, assimilated and improved when effort has been devoted to establishing the cognitive structural base in the related field. A certain level of human resource is the perquisite for the learning, and R&D departments play an important role in organizing the talents to conduct systematically collective learning.

R&D functions as a stable element of embedding new knowledge in routines of firms and transferring the tacit knowledge through the interaction with other departments. Thereby, R&D is the primary agent of organizational learning within the firms. It is not only important in innovation, but also bears social rate of return by influencing the absorptive capacity of the firms (Griffith et al. 2003). In Cohen and Levinthal's (1990) exploratory empirical investigation on absorptive capacity, R&D activity is assumed not only to generate new knowledge but also facilitate learning by building up the firms' absorptive capacity.

In developing countries where a significant technological gap with developed countries prevails in many technological fields, the main task of R&D activities is to absorb the advanced technology rather than conduct radical innovation. The essence of this argument lies in the fact that technology is not readily-made, it requires a certain level of capacity to apply it to full use. Arrow (1969)'s vivid example on jet plan suggests that after Britain sent the jet plans to America during the Second World War, it took ten months for the Americans to conform the plans to the American usage. Evenson and Westphal (1995) propose that technology transfer to relatively technologically backward countries should be assessed with the capability to make efficient use of them. The emphasis of R&D activity in developing countries is development rather than research. To be specifically, this includes absorbing the tacitness in the new knowledge, adapting it to the local condition and improving it with the combination of firm-specific routines. Therefore, the basic aim of R&D in developing countries is to absorb, adapt and exploit such as in activities like reverse engineering and minor design improvement.

By incorporating Schumpeter's (1942) framework that emphasizes the partially excludable nature of knowledge, the recent literature on income convergence suggest that catching up in developing countries could be realized by quality-augmenting innovation whose size depends on the distance from the technological frontier (Aghion and Howitt 1998; Howitt 2000). This is exactly where the R&D functions in latecomer firms should play a role. In essence, R&D activities reflect the latecomer firms' incentive to construct absorptive capacity to gain spillover effect from the higher-level of R&D activities in developed countries. Empirical works on economic growth in developing countries account total factor productivity (TFP) growth primarily to the effect of R&D spillovers from developed countries (Grossman and Helpman 1990; Coe and Helpman 1995; Grossman and Helpman 2002). Meanwhile, some scholars find evidences to support the argument that foreign R&D spillover has greater effect on firms that undertake more R&D activities, which has greatly enhance the efficiency of applying the imported technology (Jaffe 1986; Eaton et al. 1998). Griffith et al. (2004) further point out that the need and effect of R&D functions is even larger when technological gap is large.

To sum up, R&D activity matters for the latecomer firms in the way that it enhances the absorptive capacity to assimilate and apply the advanced technology, and enables the firms to efficiently use external knowledge to foster innovation. In more complex activities, the marginal effect of R&D increases when the learning efforts are demanding (Cohen and Levinthal 1990).

4.3.3 Production Experience

Productivity differs among industries (Cameron 1996, Harrigan 1999), which implies that technology in specific industries is not neutral. In fact, technology varies in terms of the complexity and difficulty to decode it. For example, technology in standardized industries and low tech industries are more codified and is easier to be attained via market or through strict network relations. On the contrary, technology that is high-end and has not yet stabilized is more complex and tacit, requiring more communication in the knowledge transfer and more interaction in the knowledge creation process. Cohen and Levinthal (1990) gave a vivid example on this point, elaborating that a unit of knowledge advance in semiconductor industry is able to yield larger performance payoffs than in steel industry. Thereby, the current technological field that the firms are actively involved in influence the return of conscious investment on absorptive capacity such as R&D activities and training.

Firm's current production experience in specific technological field confer the necessity to invest in absorptive capacity to enable the interactive learning with external partners. Meanwhile, the initial technological field, which can be defined as the technological level of the products when the firms start the business, constitutes the basic elements of the cognitive structure for further development. As absorptive capacity is actually closely related to the prior related knowledge of firms, it can be generated through direct involvement in related manufacturing operations (Rosenberg 1982). In other words, the initial production experience in specific technological field determines the initial stock of knowledge and capital. For example, firm A started with producing hard disk is endowed with higher skill as well as sophisticated machinery than firm B started with producing portable disk. Therefore, the potential for further learning is higher in firm A.

The importance of initial knowledge stock is related to the path-dependent accumulation path of firm routine. Firms in high-tech fields do not only possess higher endowment of absorptive capacity to process external complex knowledge, but also tend to form a higher expectation on the commercial value of latest technological advances (Cohen and Levinthal 1990). This phenomenon has been termed as the "lock-out" effect, in which the firms that failed to invest in certain absorptive capacity in certain technological fields tend to be discarded out of the profit room from the rapid technological opportunities in that specific field (Cohen and Levinthal 1994).

Nevertheless, the lock-out effect also occurs in the so-called high-tech industries when the technological paradigm in the industry changes dramatically in a new round of technological cycle. The Swiss watch industry, for instance, demonstrates the organizational inertia in the face of fundamental technological changes from mechanics to electronics (Glasmeier 1991). In times of technological discontinuity, there are opportunities for the technological latecomers to forge ahead as they do not have to invest a lot to unlearn the previous knowledge and routines. On the other way around, the previous technological leaders have to invest more to incorporate new skills and knowledge into the old systems of production.

The initial technological fields do not only determine the prospect of future learning owing to its path-dependent nature, but also take effect in attracting more advanced technological spillover from the global lead firms. This is because that the production experiences of firms define the opportunity of wider cooperation and alliances with other firms. Especially for the latecomer firms, the large MNCs tend to identify those firms that have recorded remarkable performance profiles in related technological fields as their strategic suppliers or even business partners (Wang and Blomstrom 1992).

4.3.4 Brief Summary

Figure 4.3 summarizes the sources of a firm's absorptive capacity, which is necessary for the use of external knowledge. These components are interrelated with each other and jointly influence on firms' incentive, aspiration and capability in assimilating, transforming and exploiting the external knowledge. Hereby, the following hypothesis can be concluded:

Hypothesis 4 The higher absorptive capacity the firms possess, the easier they can understand and communicate with external actors in the knowledge transfer process, and thus the more they tend to interact with external actors to foster innovation.

What still remains unclear in the relationship between absorptive capacity and interactive learning with external partners is its impact on the use of proximity in interactive learning activities. As absorptive capacity is embodied as combination of different components and is thus idiosyncratic among the firms, it would be fairly mechanic to analyze the optimal level of absorptive capacity. On the other

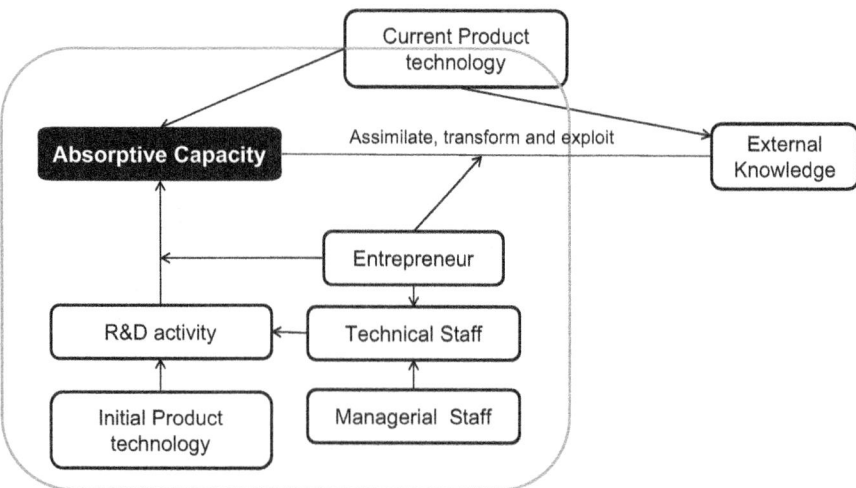

Fig. 4.3 Sources of firm absorptive capacity

way around, it would be more meaningful to consider the specific components of absorptive capacity that support the use of different proximities in order to sustain trustful and effective interactive learning activities. To be more specific, the chapter would like to take the exploratory step to investigate on the specific components of absorptive capacity that support the use of organizational proximity as well as social proximity in the interactive learning. Answers to this inquiry are able to shed light on the conscious investment on absorptive capacity if a firm is to organize the interactive learning with organizationally proximate partners and socially proximate partners.

4.4 Operationalization of Analysis

The electronics industry in the PRD, China, has been selected as the research area for this study. Because the investigation focuses on the electronics industry in an export-oriented region, geographical proximity is guaranteed due to the co-location of the firms in the same mega-urban area. In this chapter, I focus on the role of social proximity, i.e. embedded in *Guanxi* networks between individuals, as well as organizational proximity to global firms in fostering product innovation. These two forms of proximity can be addressed by conscious firm strategies, and can thus be achieved through the efforts of individual firms. In contrast, institutional proximity is not discussed, since the institutional environment is not yet stable enough for firms to rely on it, and individual firms are not able to influence the institutional setting through their own efforts in the short run.

The empirical data used to answer the research question was taken from a standardized survey of electronics firms in the PRD, Guangdong Province, China. Bellandi and Tommaso (2005) points out that the industrial development in Guangdong Province is the subtle mixture of global network, public governance and the unexplored socio-cultural contexts. In this chapter, survey data in the firm level will be used to explore the role of socio-cultural factors, i.e., social proximity in fostering innovation.

The sample and the primary questions applied in this chapter is the same as the one in Chap. 3. However, the ways to analyze the dataset differ owing to different research aims. In Chap. 3, the clustering procedure aims to reveal different firm clusters in the degree of undertaking interactive learning activities and examine the effect of interactive learning on innovation outcomes in a general term. This chapter goes further to explore the different behavioral patterns among the firms in undertaking interactive learning. As a result, it aims to identify firms capitalizing on different proximity in the interactive learning to foster innovation and how this behavioral pattern relates to the firms characteristics such as size and absorptive capacity.

Table 4.3 presented the dimension of the indicators, which related to the previous hypothesis on the use of proximity in the interactive learning.

Table 4.3 Indicators of proximity use in interactive learning

		Remarks
Obtaining New Product Ideas	Internal Efforts	*Own development of ideas*; Self Absorption and Learning through *license purchasing and reverse engineering*
	With organizationally proximate partners	Interacting with *parent companies & foreign customers*
	With organizationally distant partners	Interacting with *domestic customers, foreign customers, universities, research institutes and sales agents*
Obtaining Codified Knowledge	Internal Efforts	Self-purchasing of equipment and software
	With organizationally proximate partners	Interacting with *parent companies & foreign customers*
	With organizationally distant partners	Interacting with *domestic customers & foreign customers*
Obtaining Tacit Knowledge	With organizationally proximate partners	Receiving training and know-how from people sent *by parent company & foreign customers*
	With organizationally distant partners	Receiving training and know-how from people sent *by domestic customers and foreign customers*; Sending staff *to domestic customers or domestic lead firms, foreign customers or foreign lead firms, and universities* for training
Interaction Mode	Informal *Guanxi* Network	Interacting through *Guanxi*, for example gaining information on the reputation and capacity of innovation partners from other *business partners, relatives and friends* in the innovation process
	Active Searching	Searching for information on partners *via Internet, exhibition and sales agents* in the innovation process

For detained information on question formulation, please refer to Sect. 7.1 Appendix A: Firm Questionnaire, Part C, Question 27–30

In the questionnaire design, it is difficult to identify the interaction mode with each business partner in each specific innovation process. If so, the matrix of questionnaire would be too complex for the firms to answer. In order to ensure the success of the survey, only general information on interaction way with business partners can be identified. However, it is considered that by differentiating firm with the type such as parent companies, foreign customers, domestic customers and external knowledge institutes, information on the proximity use can be attained to some extent. As discussed in the theoretical part on the role of proximity in fostering interactive learning, firms either interact with organizationally proximate partners such as parent company or strictly controlled foreign customers, or they establish

4.5 Empirical Evidence

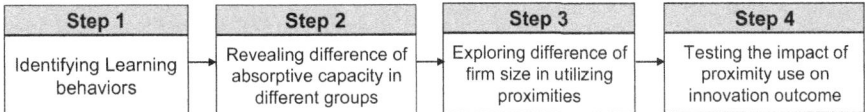

Fig. 4.4 Operationalization of analysis

social proximity with organizationally distant partners to ensure trust and understanding in the process of interactive learning. Thereby, the social proximity with domestic customers and external institutes can substitute the lack of organizational proximity in some degree. Combined with the general question on interaction mode with all business partners, insight into the degree of proximity use in innovation process can be gained.

From Figs. 3.1, 3.2 and 3.3 in Chap. 3, it is already revealed that the electronics firms in the PRD seldom resort to the parent companies for innovation ideas and tacit knowledge. The interaction with foreign firms also weighs less than that with domestic customers. Although the data shows a decreasing intensity of using organizational proximity in the whole sample, there could be a group of firms that still rely more on organizational proximity than other firm in innovation activities. Thereby, data should be processed to extract the comparative degree of using organizational proximity with parent companies and foreign customers and social proximity with other organizationally distant partners. Figure 4.4 shows the operationalization of the analysis that makes the analysis of proximity possible.

4.5 Empirical Evidence

4.5.1 Innovation Behavior of Electronics Firms

Factor analysis is firstly applied to extract the latent variables among the original variables that investigate the innovation behaviors of firms in the product innovation process. In factor analysis, factor scores of each latent variable are calculated for each case based on the regression on all variables to the latent variable. The factor scores are all standardized value, which reflects the comparative value compared to the mean of all cases. Therefore, factor analysis before the cluster analysis contributes to the aim of this chapter in the way that the difference between the variables are reduced while only difference between the cases are analyzed to bear the clustering results. Translating this methodological specificity into the empirical implication in this chapter, it means that the following clustering would base the analysis mostly on the difference among the cases in regard to the scope of interactive learning instead of the intensity of interactive learning. Moreover, factor analysis is able to explain complex phenomena in regard to the wide scope of interactive learning by extracting the main factors.

Table 4.4 Factor analysis of obtaining new product ideas (NPI)

	Factors		
	NPI_external partners	NPI_internal efforts	NPI_parent comp. & foreign customers
1) market report of sales agents	*0.78*	0.22	0.06
2) Orders from domestic customers	*0.74*	−0.09	0.10
3) Market report of university and research institute	*0.60*	0.41	0.20
4) Own idea generation	−0.03	*0.79*	−0.17
5) Reverse engineering	0.13	*0.58*	0.22
6) Purchase product licenses	0.18	*0.54*	0.51
7) Orders from parent company	0.04	0.11	*0.87*
8) Orders from foreign customers	0.46	−0.09	*0.56*
Explained variance (%)	33	15	12
Total explained variance	60%		

Tables 4.4–4.7 shows the result of factor analysis. In statistical term, the result is satisfactory because the derived factors in each group are able to explain over 60% of variance in the original sample.

In respect to the source of new product ideas (Table 4.4), three main factors are concluded, which explain 60% of the total sample variance. The three factors also bear theoretical meaning. The first factor implies that the firms interact more with the external partners such as domestic customers, universities, research institutes and sales agents to trigger new innovative ideas. These external partners bear little organizational proximity with the firms and require other proximities to support the trust-based interactive learning process. From the explained variance, the first latent factor, i.e. interacting with external partners as a way to trigger innovation ideas is the most important channel that the electronics firms apply in the PRD. The second factor have more weigh for own idea generation, reverse engineering and product license purchasing as the way to generate new innovation ideas. These activities require more input of internal resources as well as higher capability of firms to absorb or even creatively recombine the knowledge embodied in the licensed product or advanced samples. The third factor put more weighs on the interaction with parent companies and the foreign customers, which is more powerful in governance and bears closer organizational proximity. Because of the unbalanced power relation in regard to capabilities, getting new product ideas in this way is rather passive and mainly from the commands. On the other way, getting new product ideas from the domestic customers, as indicated by the first factor, might be more active and interactive due to the balanced power and capability. The assumption behind this conclusion is that local firms are latecomers so that they are more in the same level of technological capabilities and market knowledge compared to that with foreign firms.

4.5 Empirical Evidence

Table 4.5 Factor analysis of obtaining new product codified knowledge (NPCK)

	Factors		
	NPCK_customer	NPCK_parent comp.	NPCK_self purchase
1) Acquisition from domestic customer	0.86	−0.21	0.05
2) Acquisition from foreign customer	0.74	0.41	−0.03
3) Acquisition from parent company	0.00	0.94	0.04
4) others	0.02	0.04	0.99
Explained variance (%)	34	27	24
Total explained variance	85%		

In respect to getting codified knowledge in the process of product innovation (Table 4.5), three factors are derived, and they explain 85% variance of the total sample. Note that codified knowledge here refers to knowledge embodied in equipment, machinery and operational software. The first factor implies getting equipment from domestic and foreign customer. The second factor implies that firms rely more on the very close organizational proximity (parent company) to get support of required equipment in the process of product innovation. The last factor implies other ways of getting codified knowledge. In the questionnaire, firms can answer openly through which channel they actually get the equipment, and 85% of the firms answer "self-own" of "self-purchase", which bears a more internal characteristic.

In respect to getting tacit knowledge in the process of product innovation (Table 4.6), again three factors are derived, which explain 74% variance of the total sample. It should be noted that tacit knowledge here refers to technical experiences and know-how that is easier to be understood and absorbed through face-to-face interaction. The first factor distinguishes itself from the other two factors, since it implies a more active strategy of searching tacit knowledge, i.e. the firms send the employees to other firms or universities to learn technical experience that is needed to guarantee the success of innovation. The second factor implies a passive way of getting required know-how and technical experience in the product innovation process, in which the firms are being taught and instructed from the engineers sent by the domestic customer or foreign customer. Similarly, the third factor implies getting tacit knowledge mainly from the parent company. In theory, the access to tacit knowledge increases from the first factor to the last one as the organizational distance decreases.

When interacting with different actors in the above aspects in the process of product innovation, the interaction way can be either formal or informal. Firms can formally conduct active searching through exhibitions, internet or sales agents to get in touch and keep contact with the interacting actors. It is assumed that

Table 4.6 Factor analysis of obtaining new product tacit knowledge (NPTK)

	Factors		
	NPTK_active searching	NPTK_passive from customer	NPTK_passive from parent comp.
1) Engineers sent to universities	*0.85*	0.13	0.09
2) Engineers sent to domestic lead firms or customers	*0.79*	0.20	−0.11
3) Engineers sent to foreign lead firms or customers	*0.60*	0.42	0.31
4) Engineers sent by domestic customer	0.18	*0.85*	−0.13
5) Engineers sent by foreign customer	0.24	*0.78*	0.29
6) Engineers sent by parent company	0.03	0.05	*0.94*
Explained variance (%)	44	17	13
Total explained variance	74%		

interaction in this way bears an arms-to-length market relationship. Besides, they can also interact with them through the informal *Guanxi* network, such as recommendation from business partner, friends and relatives, which bears a reciprocal favor exchange and responsibility in traditional Chinese society. Table 4.7 shows that a dichotomy dimension of formal and informal way of interaction is derived, which explains 79 % of variance of the total sample. Particularly, the informal interaction way explained over half of the variance.

By means of the factor analysis, different dimensions of proximity for the interaction with different players have been identified. The small explained variances in factors concerned about the interaction with parent companies or foreign customers correspond to the results shown by Figs. 3.1, 3.2 and 3.3 in Chap. 3, indicating the low tendency of electronics firms to use organizational proximity in the interactive learning. Instead, they tend to interact with external partners beyond the organizational boundaries, for instance domestic customers, universities, research institutes and sales agents, to get new ideas and needed knowledge. Moreover, it is proven that *Guanxi* networks are the major facilitator for interaction during innovation processes. The overall result of the factor analysis is summarized in Table 4.8.

After the factor analysis, a cluster analysis uses the latent variables derived from the factor analysis to identify patterns of capitalizing on social and organizational proximity. In cluster analysis, there is rarely one single best solution. A good cluster analysis should at first use as few clusters as possible and secondly capture all

4.5 Empirical Evidence

Table 4.7 Factor analysis of interaction mode

	Factors	
	NPInteraction_informal	NPInteraction_formal searching
1) Personal contacts (recommendation from family members and friends)	0.88	−0.01
2) Business contacts (e.g. recommendation from partners)	0.73	0.31
3) Active searching (e.g. exhibitions, internet, sales agent, etc.)	0.13	0.97
Explained variance (%)	52	27
Total explained variance	79%	

statistically and empirically important clusters. I follow a four-step procedure to ensure the internal validity of the clustering result (Delmar et al 2003).

Step 1: Hierarchical clustering with Ward's method and Euclidean distances was run to assess the possible clustering results. In this step, I came with 2 to 6 cluster solutions and derived each centroid from each cluster solutions. Clustering results in this step serve as the try-out sample to theoretically assess the optimum number of clusters based on interpretability.

Step 2: I use the centroids derived in the first step to perform the K-means cluster. The results of K-means cluster serve as a hold-out sample, for which I would use to validate the results from the try-out sample.

Step 3: Hold-out sample would be compared with try-out sample with means of cross tabulation. A significant level in Lambda lower than 0.05 is considered to be able to verify the relative stability of the cluster results across samples, accepting the null hypothesis that the two samples are closely correlated. Under this confidence level (0.05) the solution stability can be assured and the clustering solution can be selected based on parsimonious interpretability.

Table 4.9 shows the results of cross tabulation. After running these two procedures, three clusters solution is shown as an internally stable and easily interpreted solution, at best explaining the innovation behaviors using different proximities in a theoretically sound and parsimonious way.

Table 4.10 demonstrates the results of our cluster analysis, which differentiates between three types of innovation behavior related to the capacity of capitalizing on social and organizational proximity in the process of product innovation. The results correspond rather well with our conceptual considerations.

- **Socially active innovator:** Firms in this group interact frequently with external partners in combination with their internal capability. With regard to obtaining codified and tacit knowledge in the product innovation process, firms of this kind tend to rely more on customers, and use the active strategy of sending people to business partners for acquiring tacit knowledge. In the interaction process with

Table 4.8 Summarization of factor analysis

		Remarks	Explained variance of each factor (%)	Total explained variance (%)
Obtaining new product ideas	NPI_external partners	Interacting with *domestic customers, universities, research institutes* and *sales agents* to gain innovation ideas	33	60
	NPI_internal efforts	Making *internal learning efforts* such as own ideas, license purchasing and reverse engineering	15	
	NPI_parent comp. & foreign customers	Relying on parent companies or foreign customers to gain innovation ideas	12	
Obtaining codified knowledge	NPCK_customer	Interacting with *foreign and domestic customers* to get codified knowledge	34	85
	NPCK_parent comp.	Interacting with *parent companies* to get codified knowledge	27	
	NPCK_self purchase	Purchase equipment self	24	
Obtaining tacit knowledge	NPTK_active learning	Sending staff *to business partners* for training	44	74
	NPTK_passive from customer	Receiving training and know-how from people sent *by domestic and foreign customers*	17	
	NPTK_passive from parent comp.	Receiving training and know-how from people sent *by parent company*	13	
Interaction mode	NPInteraction_informal	Interacting with innovation partners *within Guanxi networks*	52	79
	NPInteraction_formal searching	Interacting with innovation partners by *searching via Internet and exhibition*	27	

Table 4.9 Cross tabulation between try-out sample and hold-out sample

	Lamda value	Significance level
2-clusters solution	0.577	0.088
3-clusters solution	0.618	0.047
4-clusters solution	0.723	0.038
5-clusters solution	0.694	0.038
6-clusters solution	0.708	0.035

4.5 Empirical Evidence

Table 4.10 Cluster analysis results (Ward's method/Squared Euclidean distance)

	Socially active innovator	Organizationally dependent innovator	Lame innovator
NPI_external partner	0.54	0.25	−0.32
NPI_internal	0.52	0.07	−0.31
NPCK_customer	0.60	−0.15	−0.34
NPTK_passive from customer	0.46	0.07	−0.22
NPTK_active learning	0.58	−0.12	−0.35
NPInteraction_informal	0.60	−0.06	−0.33
NPInteraction_formal searching	0.26	−0.01	−0.17
NPI_parent comp. & foreign	−0.11	1.01	−0.12
NPCK_parent comp.	−0.38	1.96	−0.27
NPCK_self purchase	−0.17	0.12	0.10
NPTK_passive from parent comp.	−0.47	2.06	−0.16
Number	*104*	*41*	*171*

these partners, firms in this category flexibly combine formal active searching and informal networks, i.e. *Guanxi* with family members, friends and business partners. Although it is not possible to specify exactly which interaction mode is applied by the firms when interacting with each partner (because the related matrix would be too complex to be answered by the firms), it is possible to conclude indirectly that firms in this group rely on social proximity to external partners in general during the process of product innovation to a greater degree than firms in the other two clusters. They are actually socially active innovators, and social proximity is not only used as a way of acquiring codified and tacit knowledge, but also as a way of triggering new product ideas, which is a feature of capable firms in a well-functioning regional innovation system.

It is worth mentioning that although these firms are already able to extend the scope of interactive learning in the innovation process to capitalize further on social proximity, they still rely on organizational proximity to foreign customers to a certain degree in order to acquire codified and tacit knowledge. This again supports the mutual reinforcing effect of social proximity and organizational proximity. Social active innovator tend to apply mixed strategies in using proximity to facilitate interactive learning.

- **Organizationally dependent innovator:** In contrast, organizationally dependent innovators rely heavily on organizational proximity to gain access to and absorb knowledge. They turn to their parent companies to obtain codified and tacit knowledge in the process of product innovation, acting in a more passive way due to the hierarchical control. The new product ideas originate mainly from parent companies as well as from powerful foreign customers.

What is again noteworthy is that organizational dependent innovators show a certain tendency to interact with external partners to prompt product innovation, although with a lower degree than socially active innovators. However, the much lower value in informal interactions indicates that these firms are not able to capitalize on social proximity to foster innovation as their socially active counterparts. Moreover, their mode of interacting with innovative partners is not characterized by any particular feature, which indicates a more passive attitude towards product innovation compared to socially active innovators.

- **Lame innovator:** Compared to the previous two kinds of firms, lame innovators have low values for all the indicators that are related to product innovation. Lame innovators are not actively involved in triggering new ideas of innovation, nor do they strive to search for codified and tacit knowledge, which is important for positive product innovation outcome. Moreover, they are quite vague and unsettled in their ways of interacting with partners in the innovation process. In short, they are not able to interact with external players to initiate innovation and do not have the capacity to organize internal learning.

A look at the number of firms in each cluster shows that the number of lame innovators exceeds the sum of socially active and organizationally dependent innovators in our sample. This is proof to the immature absorptive capacity of most firms in the PRD to benefit from external interaction in order to trigger innovation. However, the number of socially active innovators is two times higher than the number of organizationally dependent innovators. This seems to be an indication of a maturing regional innovation system in the PRD, where some local firms are capable of benefiting from localized knowledge sources by capitalizing on informal social relations. But it also reflects the difficulty of most firms in the PRD to 'couple strategically' with global firms to upgrade their position in the value chain. By studying the relocation issue of Taiwanese Personal Computer firms, Yang (2009) also pointed out that Taiwanese firms in the PRD are less oriented towards the strategic coupling of local and global knowledge sources than their counterparts in the Yangtze River Delta.

The validation of the clustering result is further supported by external validity. If the clustering discriminates between variables not included in the clustering procedure, the clustering result is likely to represent distinct empirical categories. The clusters identified are then compared with respect to performance indicators such as sales growth, export orientation and product innovation outcomes.

Table 4.11 presents differences in sales growth, export orientation and product innovation outcomes between the clusters. Except for insignificant differences for the new product rate and functional expansion performance between each of the groups, at least one pair of groups differs significantly from another. This again validates the explanatory power of the three clusters.

4.5 Empirical Evidence

Table 4.11 Performance of different innovating groups (T test of samples)

		Group 1	Group 2		Group 1	Group 3		Group 2	Group 3
Sales growth in first half of 2009	Mean	20.9	−4.23	Mean	20.9	24.4	Mean	−4.23	24.4
	Sig.[a]	0.036		Sig.	0.601		Sig.	0.023	
Foreign market share		Group 1	Group 2		Group 1	Group 3		Group 2	Group 3
	Mean	36.3	55.4	Mean	36.3	44.4	Mean	55.4	44.4
	Sig.	0.003		Sig.	0.064		Sig.	0.084	
New product rate		Group 1	Group 2		Group 1	Group 3		Group 2	Group 3
	Mean	36.9	33.3	Mean	36.9	37.6	Mean	33.3	37.6
	Sig.	0.422		Sig.	0.815		Sig.	0.360	
Quality improvement		Group 1	Group 2		Group 1	Group 3		Group 2	Group 3
	Mean	4.24	4.27	Mean	4.24	4.01	Mean	4.27	4.01
	Sig.	0.844		Sig.	0.056		Sig.	0.111	
Function expansion		Group 1	Group 2		Group 1	Group 3		Group 2	Group 3
	Mean	3.84	3.80	Mean	3.84	3.62	Mean	3.80	3.62
	Sig.	0.826		Sig.	0.125		Sig.	0.388	
Product upgrading		Group 1	Group 2		Group 1	Group 3		Group 2	Group 3
	Mean	3.79	3.95	Mean	3.79	3.52	Mean	3.95	3.52
	Sig.	0.442		Sig.	0.087		Sig.	0.030	

Group 1 Socially active innovator, *Group 2* Organizationally dependent innovator, *Group 3* Lame innovator
[a] Significance level of the difference between the mean value of the comparing groups

- Differences between organizationally dependent innovators and the other two groups

  Almost 58% of the organizationally dependent innovators in our sample involve foreign ownership, while the share is only 30% among socially active innovators and 38% among lame innovators. This indicates that organizationally dependent innovators are more closely linked to the global production network. This is further substantiated by their outstanding export performance compared to the other two groups. However, sales growth after the financial crisis is negative and significantly lower than that of socially active innovators and even lame innovators. This suggests that the loose social embeddedness with local business partners led to a highly vulnerable and externally dependent mode of upgrading. With regard to product innovation, organizationally dependent innovators are more able to upgrade the product category (e.g. from mp3 to mp4) by integrating within the global value chain compared to the lame innovators. However, they do

not outperform the lame innovators in other aspects of product innovation, which suggests that depending solely on global production networks narrows the scope of product innovation.
- Differences between socially active innovators and lame innovators

    The two groups represent a high share of domestic-oriented firms, especially the socially active innovators, whose export share is only around 36%. They are all able to reach moderate sales growth even in the face of the crisis. However, socially active innovators outperform lame innovators in many aspects of product innovation, such as quality improvement and product upgrading facilitated by the capacity to take advantage of informal social relations as well as some degree of organizational proximity to foreign customers.

Lame innovators represent the conventional producers in the PRD. They are able to respond to market needs with low-cost and flexible production by taking advantage of informal relations with family members and friends. However, their lower absorptive capacity restricts them to use informal social relations to foster innovation and upgrading. These firms represent the primary bottleneck for a shift in simple production towards innovation and upgrading in the PRD.

4.5.2 Absorptive Capacity and Learning Behaviors

After the identification of different learning behaviors, the analysis aims to further analyze the relationship between the internal absorptive capacity and the choice of proximity use. The hypothesis discussed in the Sect. 4.3 suggests that certain level of absorptive capacity is the basic perquisite of interactive learning, which helps in effective communication in the knowledge transfer process as well as absorbing and exploiting the external knowledge. Sources of absorptive capacity are defined in this study as human capital, R&D activities and production experience.

According to hypothesis four, internal absorptive capacity defines the capacity of firms to understand and communicate with external actors in the knowledge transfer process, as well as possibility of using proximities to foster innovation. Table 4.12 demonstrates the specific indicators for internal absorptive capacity.

The construction of the indicators of absorptive capacity is based on the theoretical discussion in Sect. 4.3. As for the human capital, educational level of the technical staff and managerial staff serves as the base for the rate of human capital accumulation. With the training efforts, the human capital accumulates as the capacity of absorbing and exploiting new knowledge to the need of the firm development is enhanced. The training effort is formulated as the frequency of training: one indicates only once upon recruitment, two indicates training that is more often but on an irregular basis, and three indicates regular and systematic training. The percentage of technological or managerial staff that is bachelor degree or above would be multiplied by their respective training frequency to signify the level of human capital to absorb, apply and exploit knowledge. In addition, staff training expenses, CEO education and CEO work experience would be measured.

4.5 Empirical Evidence

Table 4.12 Indicators of absorptive capacity

	Indicators	Description
Human Resource	Level of technical staff	Percentage of technical staff that have bachelor degree or above *multiplied by* training frequency
	Level of managerial staff	Percentage of managerial staff that have bachelor degree or above *multiplied by* training frequency
	Staff training expenses	Expenses on staff training in the year 2007
	CEO education	1 as below bachelor degree
		2 as bachelor degree
		3 as graduate degree (master or doctor)
		4 as bachelor or above combined with overseas experience
	CEO work experience	0 as no former working experience
		1 as private sector working experience only
		2 as public sector working experience (might involve in private sector once)
		3 as overseas working experience
R&D activities	Technology center (R&D Center)	1 as having technology center (R&D center), 0 as not
	Design capability	1 as having product design capability, 0 as not
	Development capability	1 as having product development capability, 0 as not
Production experience	Current production experience	Defined according to International Standard Industrial Classification of all Economic Activities, Rev 3[1], 1 as initial product embodying high technology, 0 as initial product embodying low and medium technology
	Initial production experience	Defined according to International Standard Industrial Classification of all Economic Activities, Rev 3[1], 1 as current product embodying high technology, 0 as current product embodying low and medium technology

Specific classification of products into the different levels could be referred to Sect. 7.3 Appendix C

In the next step, the difference of these capabilities among the three different innovating groups is tested, which provide answers to the question: Which components of absorptive capacity enable a firm to conduct extra-learning using either organizational proximity or social proximity in product innovation process?

In SPSS procedure, I firstly apply independent samples T test to examine whether there is significant difference between the mean values of absorptive capacity among each pairs of innovating groups. The significance level indicate that the

Table 4.13 Human capital of different innovating groups (T test of samples' mean)

Level of technical staff (%)		Group 1	Group 2		Group 1	Group 3		Group 2	Group 3
	Mean	75.8	90.4	Mean	75.8	69.6	Mean	90.4	69.6
	Sig.[a]	0.337		Sig.	0.552		Sig.	0.144	
Level of managerial staff (%)		Group 1	Group 2		Group 1	Group 3		Group 2	Group 3
	Mean	*81.6*	*110.0*	Mean	81.6	83.4	Mean	*110.0*	*83.4*
	Sig.	*0.064*		Sig.	0.873		Sig.	*0.077*	
Training expenses in 2007 (Unit: Yuan)		Group 1	Group 2		Group 1	Group 3		Group 2	Group 3
	Mean	*93,765*	*755,258*	Mean	93,765	146,748	Mean	*755,258*	*146,748*
	Sig.	*0.047*		Sig.	0.332		Sig.	*0.067*	
CEO education		Group 1	Group 2		Group 1	Group 3		Group 2	Group 3
	Mean	*2.54*	*2.88*	Mean	2.54	2.46	Mean	*2.88*	*2.46*
	Sig.	*0.075*		Sig.	0.500		Sig.	*0.019*	

Group 1 Socially active innovator, *Group 2* Organizationally dependent innovator, *Group 3* Lame innovator
[a] Significance level of the difference between the mean value of the comparing groups

difference between the mean value of the comparing groups is significantly (at least at 90 % level) greater than 0.

Table 4.13 shows the results of independent samples T test of each 2 groups for human capital. The insignificant difference of all indicators of human capital between socially active innovator and lame innovators demonstrate that human capital is not the determinant factor for the use of social proximity to undertake interactive learning.

What stands out from the result in Table 4.13 is the significant higher level of managerial staff level, CEO education and staff training expenses for organizationally dependent innovator compared to the socially active innovator and lame innovator. This might be attributed to the capability of high-level managerial staff and entrepreneurs to enable "strategic coupling" with global firms. Thereby, when sophistication of global value chain allows co-evolvement and upgrading of suppliers and subsidiaries in latecomer countries, the global lead firms tend to identify firms with higher human capital as the strategic partners.

From Table 4.14, it is found that firms that undertake interactive learning have a slightly higher tendency to organize R&D activities than firms that do not undertake interactive learning (lame innovator), but the differences are not in a significant level.

Cohen and Levinthal (1990)'s finding on R&D's function in creating and exploiting new knowledge is not supported with the results in Table 4.14. The marginal role of R&D activities in contributing to the development of absorptive capacity among the electronics firms in the PRD can be explained from two respects. Firstly, the overall stock of knowledge generated by systematic accumulation of R&D activities might be too small to generate the absorptive capability that is required

4.5 Empirical Evidence

Table 4.14 R&D activities of different innovating groups (T test of samples' mean)

Technology center (R&D Center)		Group 1	Group 2		Group 1	Group 3		Group 2	Group 3
	Mean	0.66	0.68	Mean	0.66	0.62	Mean	0.68	0.62
	Sig.[a]	0.818		Sig.	0.430		Sig.	0.422	
Design capability		Group 1	Group 2		Group 1	Group 3		Group 2	Group 3
	Mean	0.77	0.76	Mean	0.77	0.75	Mean	0.76	0.75
	Sig.	0.868		Sig.	0.700		Sig.	0.921	
Development capability		Group 1	Group 2		Group 1	Group 3		Group 2	Group 3
	Mean	0.79	0.73	Mean	0.79	0.70	Mean	0.73	0.70
	Sig.	0.466		Sig.	0.106		Sig.	0.707	

Group 1 Socially active innovator, *Group 2* Organizationally dependent innovator, *Group 3* Lame innovator

[a] Significance level of the difference between the mean value of the comparing groups

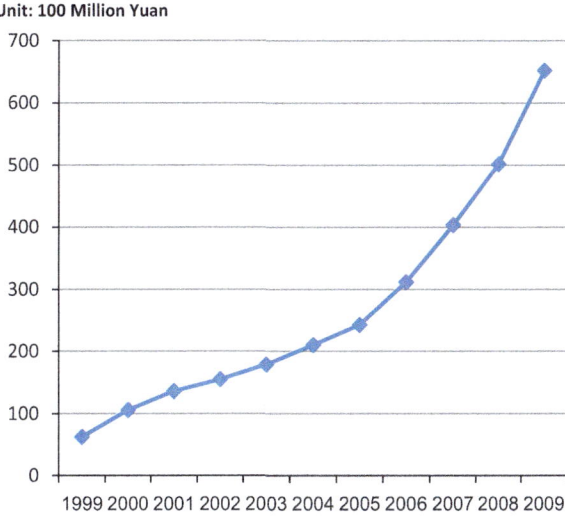

Fig. 4.5 R&D expenses in Guangdong Province (1999–2009). (Data source: CASTED 1999–2009, annual)

for effective knowledge exchange with the external partners. Although for the survey sample, the average durations of firms that have been possessing the design capability and development capability are 9.6 years and 8.9 years respectively, the statistical data in Guangdong province shows that R&D expenses just pick its accelerating rate after 2005 (Fig. 4.5). Secondly, R&D activities are not strategically organized so that knowledge accumulation is not continual and systemic to guarantee the construction of sufficient absorptive capacity. Moreover, as demonstrated by Table 4.15, the R&D expenses in Guangdong Province is extremely concentrated in investment in test and development, which includes less reflection on the long-term

Table 4.15 National comparison of technological indicators (2008). (Data sources: CSSB 2009; GPBS 2009)

	National average (%)	Guangdong (%)
R&D expense (percentage in GDP)	1.5	1.4
# Investment in Basic Research (%)	5	1.4
# Investment in Application Research (%)	12.5	1.6
# Investment in Test & Development (%)	82.5	97

Table 4.16 Production experience of different innovating groups (T test of samples' mean)

Current production experience		Group 1	Group 2		Group 1	Group 3		Group 2	Group 3
	Mean	0.19	0.24	Mean	*0.19*	*0.10*	Mean	*0.24*	*0.10*
	Sig.[a]	0.493		Sig.	*0.041*		Sig.	*0.049*	
Initial production experience		Group 1	Group 2		Group 1	Group 3		Group 2	Group 3
	Mean	0.19	0.20	Mean	*0.19*	*0.07*	Mean	*0.20*	*0.07*
	Sig.	0.853		Sig.	*0.009*		Sig.	*0.060*	

Group 1 Socially active innovator, *Group 2* Organizationally dependent innovator, *Group 3* Lame innovator

[a] Significance level of the difference between the mean value of the comparing groups

strategy of technological capability accumulation. In the analysis by Cohen and Levinthal (1990), R&D refers to more basic research activities that are able to prepare the firms with general background knowledge to exploit new scientific knowledge.

Finally, it can be detected that for the survey sample of electronics firms in the PRD, China, only the production experience differentiate the firms that undertake interactive learning (socially active innovator and organizationally dependent innovator) from the firms that do not (lame innovator). Table 4.16 shows that firms that either capitalize on social proximity or organizational proximity to foster innovation have a significantly higher percentage of previous experiences and current practices in high-tech fields than the lame innovators. This indicates that for the electronics firms in the PRD, interactive learning is practiced more by the firms currently and once involved in high-tech technological fields owing to their contribution to the absorptive capacity.

As the independent sample t-test mainly applies to comparison of mean values, chi-square is run again for categorical data in human capital and production experience to validate the robustness of the results. Especially for the CEO work experience, there is no ordered scale in the definition, and distribution test would be more appropriate than mean test.

Chi-square test for the frequency distribution of CEO education among three innovating firms shows that CEOs in organizationally dependent firms tend to have more overseas educational background, albeit the difference with the other two groups is not in a significant level (Table 4.17).

4.5 Empirical Evidence

Table 4.17 Distribution of CEO education among the innovating groups

		Innovating groups			Total
		Socially active innovator	Organizationally dependent innovator	Lame innovator	
CEO education	Below bachelor	12 (12%)	5 (12%)	25 (15%)	42 (14%)
	Bachelor	45 (45%)	11 (27%)	74 (45%)	130 (42%)
	Graduate degree	21 (21%)	9 (22%)	31 (19%)	61 (20%)
	Bachelor or above combined with overseas experience	23 (23%)	16 (39%)	35 (21%)	74 (24%)
Total		101	41	165	307

$\chi^2 = 7.92, p = 0.244$

Table 4.18 further corroborates the result of Table 4.17 and shows that organizationally dependent innovators tend to have more CEOs with overseas working experience. For the successful strategic coupling process with global partners, the network capital and market knowledge that the CEO has accumulated in the working period account more than the knowledge that they learned in the campus. The result corresponds to Saxenian and Hsu's (2001) work on "technological community", in which the Taiwanese entrepreneurs with US educating and working experience play important role in technology transfer and inter-firm collaborations between Silicon Valley and Hsinchu High-tech Park.

Table 4.18 Distribution of CEO working experience among the innovating Groups

		Innovating groups			Total
		Socially active innovator	Organizationally dependent innovator	Lame innovator	
CEO working experience	No former working experience	11 (11%)	11 (27%)	24 (15%)	46 (15%)
	Domestic private sector	49 (48%)	5 (12%)	54 (32%)	108 (35%)
	Domestic public sector	26 (25%)	7 (18%)	44 (26%)	77 (25%)
	Overseas working experience	17 (16%)	17 (43%)	45 (27%)	79 (25%)
Total		103	40	167	310

$\chi^2 = 25.341, p = 0.000$

Table 4.19 Distribution of current production experience among the innovating groups

		Innovating groups			Total
		Socially active innovator	Organizationally dependent innovator	Lame innovator	
Current production experience	Low and medium tech	84 (82%)	31 (76%)	154 (90%)	269 (85%)
	High tech	20 (18%)	10 (24%)	17 (10%)	47 (15%)
Total		104	41	171	316

$\chi^2 = 7.778, p = 0.020$

The significant level of Chi-square test in Table 4.19 upholds the results in Table 4.16. Again, the more complex components in high-tech products require more interactive process between users and producers and also with other organizations than the standardized and low-tech products. Thereby, as demonstrated by Table 4.19, the socially active innovators and organizationally dependent innovators have higher percentage of current production experiences in high-tech fields. Moreover, socially active innovators and organizationally dependent innovators have significantly higher share of high-tech endowed firms compared to the lame innovator group and even the whole sample (Table 4.20). As discussed before, the initial production experience in high-tech fields prepares the firms to undertake interactive learning with sharable knowledge with other business partners. Overall, this result supports the fourth hypothesis that higher level of absorptive capacity boosts the interactive learning activities, despite the fact that only production experience in high-tech fields serves as the valid component of absorptive capacity.

All together, the results suggest that firms that undertake interactive learning differ from the lame innovator significantly in terms of initial and current production experience in high-tech fields. Firms that have managerial staff with high education background combining with frequent training are better at using organizational proximity to foster product innovation. Moreover, it is interesting to find that CEO with

Table 4.20 Distribution of initial production experience among the innovating groups

		Innovating groups			Total
		Socially active innovator	Organizationally dependent innovator	Lame innovator	
Initial production experience	Low and medium tech	83 (81%)	32 (80%)	158 (93%)	273 (88%)
	High tech	19 (19%)	8 (20%)	12 (7%)	39 (12%)
Total		102	40	170	312

$\chi^2 = 10.160, p = 0.006$

4.5 Empirical Evidence

overseas background has a large share in organizationally dependent firm group, which indicates their role in bridging the local-global interaction in technological and market fields. OECD (2005) also points out the role of entrepreneur and their attitudes towards innovation deserve further investigation in the context of latecomer countries. Nevertheless, this study fails to confirm the role of R&D activities in shaping the sufficient absorptive capacity for interactive learning, which might be attributed to less details on the nature of R&D activities with the present indicators.

4.5.3 The SMEs' Use of Proximity

In this section, chi-square test will be applied to explore the significance of the difference between large and SMEs in using proximity in the product innovation process, aiming to provide answers to the question: What is the difference of SMEs in terms of using proximity compared to large firms?

Firstly, the overall pattern of innovating behaviors is observed by chi-square test (Table 4.21). It is shown that firm size does significantly influence the innovating behaviors. Small and medium sized firms normally lack the resources and capability to undertake external learning activities. Moreover, if small and medium sized firms are to undertake external learning activities, they tend to interact with the business partners through the use of social proximity to gain reliable information and support.

Concentrating further on the firms that apply either organizational proximity or social proximity in the product innovation process, Table 4.22 intensify the pattern on the tendency of small and medium sized firms to apply social proximity in the interactive learning process. The empirical result here supports the third hypothesis that small and medium sized firms tend to use more social proximity than large firms due to limit of internal resources and capabilities.

In Sect. 4.5.2, it is concluded that higher level of human capital gives rise to preference on organizational proximity over social proximity in the interactive learning activities. Furthermore, it has been pointed out in Sect. 4.2.3 that the lack of pecuniary resources of SMEs leads to the underinvestment in skill training, which results

Table 4.21 Difference of innovating behaviors between large firms and SMEs

Firm size[a]	Socially active innovator	Organizationally dependent innovator	Lame innovator	Total
Small and medium sized firms	99 (34%)	29 (10%)	161 (56%)	289
Large firms	5 (19%)	11 (42%)	10 (39%)	26
Total	104 (33%)	40 (13%)	171 (54%)	315
$\chi^2 = 22.504, p = 0.000$				

[a] According to Chinese statistical standard on firm size in terms of sale, firms that have with no less 300 million Yuan sales and no less than 2000 employee are assigned as large firms

Table 4.22 Difference of proximity use between large firms and SMEs

Firm size[a]	Socially active innovator	Organizationally dependent innovator	Total
Small and medium sized firms	99 (77%)	29 (23%)	128
Large firms	5 (31%)	11 (69%)	16
Total	104 (72%)	40 (28%)	144

$\chi^2 = 15.062, p = 0.000$

[a] According to Chinese statistical standard on firm size in terms of sale, firms that have with no less 300 million Yuan sales and no less than 2000 employee are assigned as large firms

Table 4.23 Difference of human capital between large firms and SMEs (T-test)

Level of technical staff (%)		Small and medium sized firms	Large firms
	Mean	70.7	137.2
	Sig.[a]	0.000	
Level of managerial staff (%)		Small and medium sized firms	Large firms
	Mean	84.7	121.8
	Sig.	0.024	
Staff training expenses (Unit: Yuan)		Small and medium sized firms	Large firms
	Mean	149428	4332045
	Sig.	0.260	

[a] Significance level of the difference between the mean values of the comparing groups

in the management incapability to internalize transactions within the firm boundary to avoid uncertainty.

Table 4.23 shows the difference between small and medium-sized firms and large firms in terms of human capital. Overall, one can see that small and medium sized firms own much less highly educated technical staff and managerial staff compared to the large firms. Furthermore, the investment in staff training is smaller in SMEs than large firms, although not in a significant level in the survey sample.

With regard to educational level of entrepreneurs, SMEs' CEOs concentrate more in the lower end of educational level compared to the large firms (Table 4.24). Nevertheless, it is worth mentioning that SME's CEOs are more experienced than large firms' CEOs in the domestic business sector (Table 4.25). The *Guanxi* networks with a wide range of business partners that the SME's CEOs have established in the past working experience might be one of the important factors for their orientation towards the use of social proximity in the product innovation process. In contrast, the richer work experience of large firm's CEOs in overseas business sector assist them in the interaction with global partners within organizational proximity.

Overall, the use of social proximity in product innovation by small and medium sized firms in the PRD signifies the shaping of reciprocal and dynamic innovation

4.5 Empirical Evidence

Table 4.24 Difference of CEO education between large firms and SMEs

		Small and medium sized firms	Large firms	Total
CEO education	Below bachelor	50 (16%)	2 (7%)	52 (15%)
	Bachelor	140 (44%)	4 (14%)	144 (42%)
	Graduate degree	60 (19%)	9 (31%)	69 (20%)
	Bachelor or above combined with overseas experience	67 (21%)	14 (48%)	81 (23%)
Total		317	29	346

$\chi^2 = 17.594, p = 0.001$

Table 4.25 Difference of CEO work experience between large firms and SMEs

		Small and medium sized firms	Large firms	Total
CEO working experience	No former working experience	42 (13%)	7 (24%)	49 (14%)
	Domestic private sector	120 (37%)	5 (17%)	125 (36%)
	Domestic public sector	79 (25%)	6 (21%)	85 (24%)
	Overseas working experience	80 (25%)	11 (38%)	91 (26%)
Total		321	29	350

$\chi^2 = 7.242, p = 0.065$

synergy among the clustered SMEs. However, the smaller share of interactive learners in the small and medium sized firm groups (again refer to Table 4.21) narrows the perspective of growing stock and further exploitation of complementary knowledge at the regional scale.

4.5.4 *Impact of the Use of Proximity on Product Innovation Outcome*

Forbes and Wield (2002) show that the firms' performance is determined by dynamic interaction of three aspects: the endowments of the firms, the channel of acquiring external knowledge and the learning efforts. Following this logic and the above theoretical discussion, a regression analysis is applied to explore the exact relationship of the proximities and product innovation outcome by controlling for firm-specific characteristics such as size, ownership, age and internal absorptive capacity.

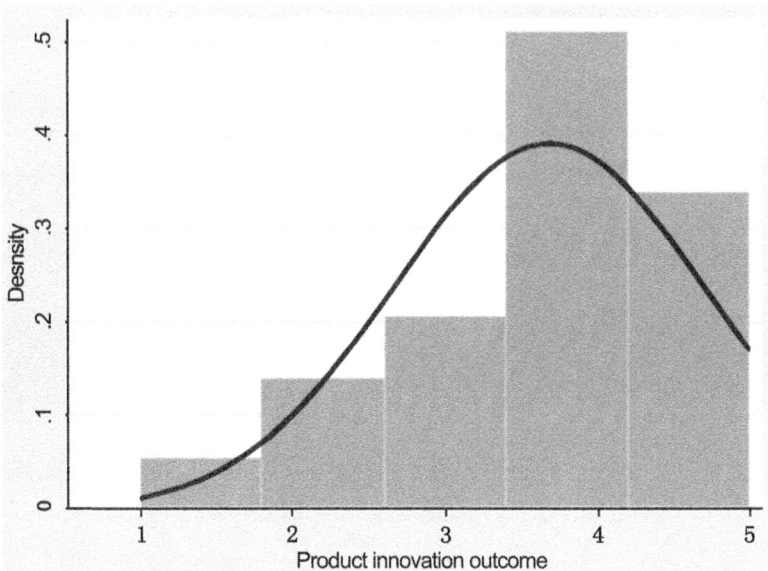

Fig. 4.6 Histogram distribution of innovation outcome on five-point likert scale

Dependent variable The dependent variable in the regression model is product innovation outcome. For survey data, especially in developing countries, it is always difficult to obtain an exact and objective measurement of new products that is reliable and comparable. Therefore, subjective measurement is taken, in which the firms are asked to evaluate the degree of improvement on product function expansion and product categories upgrading (on a scale of 1 to 5 with increasing degrees of improvement). The dependent variable in the regression is the average score of these two items.

A shortcoming of this variable is that it has a bounded value by 5. The problem here is that it is based on a subjective evaluation, and those firms that marked the same score might not be completely similar in their achievement. Figure 4.6 shows the distribution of the composite score of innovation outcome. The censoring of the data can be clearly seen, since there are far more cases with scores of 4 to 5, which can be expected in questionnaire answers because the firms all attempt to make a good impression. With this particular issue of censored data, ordinary least squares (OLS) regression provides inconsistent estimates of the parameters (Long 1997). Therefore, a tobit regression is applied which is unaffected by this issue.

Independent variables The independent variables applied in the tobit regression on innovation outcome is demonstrated in Table 4.26.

Independent variables in Table 4.26 mainly concerns three sets of indicators:

1. Firm Characteristics
   Firm characteristics such as firm size, firm ownership and firm age are applied to control the variations in the regression.

4.5 Empirical Evidence

Table 4.26 Independent variables in product innovation regression

Independent variables	Description
Firm size	Defined according to Chinese firm size standard, 1 as large firms with no less 300 million Yuan sales and no less than 2,000 employee, otherwise as small and medium-sized with the value of 0
Firm ownership	1 as firms with foreign participation (wholly owned or joint venture), 0 as firms with 100% domestic participation
Firm age	Years since establishment of the firm
Skill level of technical staff	Percentage of technical staff that have bachelor degree or above *multiplied by* training frequency
Skill level of managerial staff	Percentage of managerial staff that have bachelor degree or above *multiplied by* training frequency
CEO education	1 as CEO below bachelor degree
	2 as CEO with bachelor degree
	3 as CEO with graduate degree (master or doctor)
	4 as CEO with bachelor or above combined with overseas experience
Development capability	1 as having product development capability, 0 as not
Initial production experience	Defined according to International Standard Industrial Classification of all Economic Activities, Rev 3[1], 1 as producing low-tech products when starting business, 2 as producing medium-tech products when starting business; 3 as producing high-tech products when starting business
Innovation behavior	Defined by the cluster analysis in the next part; included in the model as a series of dummy variables

Specific classification of products into the different levels could be referred to Sect. 7.3 Appendix C

2. Absorptive capacity

- Human resource: The level of technical staff and managerial staff is applied as the proxy for human capital. Besides, the regression also applies the educational background of the CEO as an influencing factor on innovation outcomes. Leibenstein (1968) points out that in imperfect factor markets, entrepreneurs tend to carry out many activities for the survival and growth of the enterprise by themselves, such as searching and evaluating economic opportunities, taking ultimate responsibility for technical absorption and management, as well as marshaling financial resources. In our sample, 92% of the firms are small and medium sized and about 80% of employees are completely involved in production. Owing to poor endowment in resources and skill, entrepreneurs shoulder most of the responsibilities in the process of product innovation. As proxy of CEO work experiences does not take an ordered nature as the CEO education, it is not included in the regression.
- R&D activities: In order to avoid the issue of collinearity, only development capability enters into the regression function as a proxy of the presence of R&D activities.

- Initial production experiences: In the regression analysis, initial production experience defined by technological level of the primary product is further classified into three levels, i.e. low tech, medium tech and high tech. This categorical variable would be included in the model as a series of dummy variables. As current production experience displays high correlation (0.824) with the initial one, it is not included in the regression.

3. Innovation behavior: Innovation behavior which is defined by the cluster analysis before is the crux of investigation in the regression model. Again, this categorical variable would be included in the model as a series of dummy variables in the tobit regression.

Table 4.27 shows the results of the tobit regression with innovation outcome as the dependent variable and innovation behavior and other control variables as independent variables. The results of the cluster analysis are used to define the innovation behavior as: (1) socially active innovators, (2) organizationally dependent innovators and (3) lame innovators.

The chi-square likelihood ratio has a p-value of 0.009, which tells us that the model as a whole fits significantly better than an empty model. Moreover, the distribution of the residuals obey the normal rule, which indicates that heterokedastic issue, that might tortures the results of tobit model, does not exist (Fig. 4.7).

Among the variables of absorptive capacity, CEO education, development capabilities and production experience jointly constitute the primary elements of internal absorptive capacity to foster innovation. For the small and medium sized firms, the CEO acts as a gatekeeper for choosing technologies, new market opportunities and business networks. Furthermore, firms that are able to develop product on their own have better innovation outcomes. If firms initially produced high-tech products, i.e. they have accumulated production experience and related capabilities in the high-tech fields, they tend to perform better in product innovation than firms starting with low-tech production. For medium-tech endowed firms, this effect is smaller and insignificant. Lastly, it should be noted that foreign participated firms (wholly-owned or joint venture) have worse innovation outcomes than domestic firms if all other variables are held constant. This result also supports the argument in Chap. 3 that foreign firms are not active in incremental product innovation activities. Their focus might be more on high-end R&D activities or patenting. As the survey fails to identify the R&D activities in a meaningful way due to the lack of investigation on detailed composition and quality of R&D activities, it should become the focus of future research.

The main focus of the research question is the impact of the use of proximity on product innovation outcome. If control variables for firm characteristics and absorptive capacity are all held at a constant level in the model, socially active innovators possess a better product innovation outcome than lame innovators in a significant level of 0.02, while organizationally dependent innovators do not outperform the lame innovator in a significant way. This verifies the third hypothesis that social proximity is an asset that firms are able to capitalize on in complex innovation

4.5 Empirical Evidence

Table 4.27 Tobit regression on innovation outcome

Independent variables		Product Innovation outcome[a] (Average score of evaluation)
Innovation behavior	Organizationally dependent vs. lame[b]	0.23 (0.234)
	Socially active vs. lame[b]	0.37** (0.170)
	Organizationally dependent vs. socially active[c]	−0.15 (0.246)
	Overall effect[d]	—*
Firm size		−0.12 (0.293)
Firm ownership		−0.26** (0.127)
Firm age		0.005 (0.010)
Skill level of technical staff		0.0006 (0.001)
Skill level of managerial staff		0.0008 (0.001)
CEO education		0.16** (0.065)
Development capability		0.52*** (0.188)
Initial production experience	Medium tech vs. low tech[e]	0.19 (0.174)
	High tech vs. low tech[e]	0.54** (0.251)
	Overall effect[d]	*
Prob > chi2		0.0006
Pseudo R square		0.047
Number of Observations		233

[a] Product innovation outcome refers to improvement in product quality, product function and product categorical upgrading
[b] Lame innovator as the default group, which means lame innovator as 0, the other as 1
[c] Socially active innovator as the default group, which means socially active innovator as 0, the other as 1
[d] T test of whether the overall effect of the categorical variable is statistically significant
[e] Initial product as low tech as the default group, which means low tech as 0, the other as 1
Standard errors in parentheses; *$p<0.10$; **$p<0.05$; ***$p<0.01$

processes. With the development of local capabilities in the PRD after 30 years of industrialization, firms are gradually accumulating the capacity to capitalize on social proximity to foster product innovation and upgrading. Nevertheless, it also suggests that firms that apply the strategies of capitalizing on organizational proximity to foster innovation encounter difficulties in achieving satisfied innovation

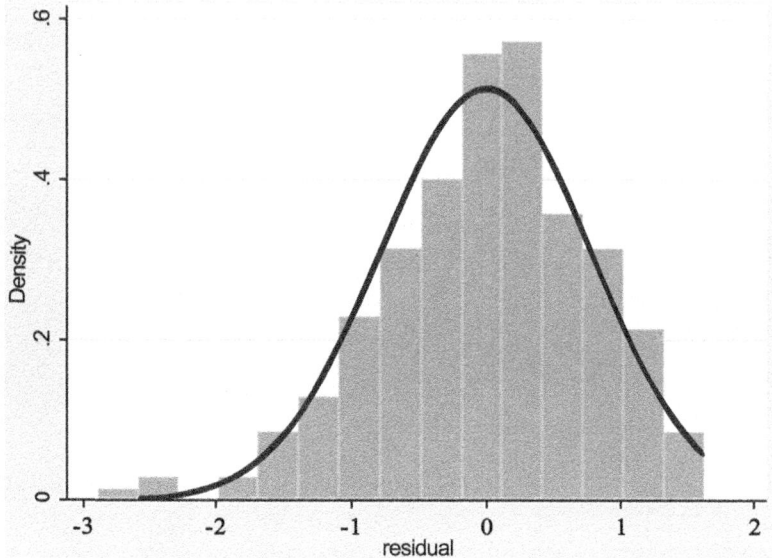

Fig. 4.7 Histogram distribution of model residuals

outcomes, which corresponds to the comparison of performance between these two groups as demonstrated in Table 4.11. The limited potential for upgrading the position in the value chain is revealed for organizationally dependent innovators, supporting the first hypothesis that is proposed in Sect. 4.2.

Nevertheless, it is necessary to cautiously examine the magnitude of improvement by applying social proximity in interactive learning. The coefficient demonstrates 0.37 degree of improvement on the average score of evaluation on production function expansion and category upgrading. To put it into practical interpretation, it means that applying social proximity in interactive learning promotes the innovation outcome either in function expansion or category upgrading by nearly one degree (e.g. from not significant to a little significant or from significant to very significant). In short, the achievement made by applying social proximity compared to applying nothing is rather small. Moreover, socially active innovators, which interact with domestic customers and other knowledge institutions in the process of product innovation, do not differ significantly from organizationally dependent innovators in terms of product innovation outcome. Even though organizationally dependent innovators were hit harder by the recent slump in global demand than socially active innovators, their innovation outcome does not differ in a substantial and significant manner compared to the socially active innovators.

This result implies an intriguing feature of the recent development stage of the regional innovation system in the PRD. Although socially active firms are emerging in the region, which altogether increases dynamic innovative synergies on the local scale, their capacity to transform fully this social asset into a high innovation outcome is not yet sufficient. This underpins the instability of innovative synergies in emerging regions where small achievements are not sufficient to compensate for the risk and cost related to innovation activities. It might be attributed to the fact that trust building needs time, especially in innovation activities that are highly complex and risky and involve high level of spillover effect. All in all, a regional innovation system is just burgeoning in the PRD, and it calls upon a stable and efficient governance infrastructure to be in place to strengthen and stabilize the interactive learning in the business sector.

4.6 Discussion and Conclusion

Proximity is a direct and simple concept dealing with the issue of learning and innovation. As Massard and Mehier (2009) suggest, it provides the measurement of accessibility other than the concept of externality as just being there. Relational space based on rules, contract and informal social interaction has been taken into comprehensive consideration.

The fact that the local firms are interested and able to capitalize on social proximity to foster innovation signifies the maturing of a regional innovation system. Moreover, the use of organizational proximity feeds dynamism into the local production system as a way to avoid negative lock-in effect. In the context of China, where low-cost is the common strategy and innovation capability is doubted, this chapter firstly gives the theoretical implication on the role of proximity in fostering innovation activities when sufficient absorptive capacity is ensured.

By examining the questionnaire data collected for the electronics industry in the PRD, China, the following trends are captured in the electronics cluster. First, as organizational proximity is taking on its limitation in respect to innovation, the electronics firms have extended the use of social proximity from low-cost production activities to undertaking interactive learning in the product innovation process. Despite the formation of a group of socially active firms, the effect of social proximity in fostering fruitful interactive learning is still limited. Second, current practice and past experience in high-tech fields have been identified for the electronics firms in the PRD as the important elements in shaping the absorptive capacity to enable the effective communication with external partners. Meanwhile, higher level of human capital such as highly educated and trained managerial staff and CEOs with overseas background is able to facilitate the interactive learning organized within organizational proximity than surpasses geographical boundary. Third, small and medium sized firms are more obliged as well as interested in using social proximity than large firms due to the constraint of financial and human resources.

The line of thinking, that social capital is an important asset for organizing interactive learning, is well revisited by the institutional and cultural turn in many schools of contemporary economic geography. In new growth theory, productive new ideas are endogenously shaped by institutional contexts (Romer 1986). The approach of innovation systems proposes that social capital induces widely spread interactive learning in the whole economy, hence creating more net wealth (Lundvall 2005). Likewise, the new institutionalism embraces again the context-dependent epistemology, considering the possibility that various social institutions in places determine the evolution of economic landscape (Clark et al. 2003). As demonstrated by the empirical investigation in this chapter, the informal *Guanxi* networks in the Chinese context are important social assets that the firms can take advantage of in ensuring effective interactive learning.

Ever since the global recession, governments at different spatial levels in the PRD feel that the strategy of low-cost production is losing its competitive edge and start to promote industrial upgrading and innovation. Empirical evidences in the chapter point out that industrial upgrading in high-tech fields should become the policy-focus because it is the precondition of active interactive learning and the formation of a dynamic regional innovation system. It is intriguing to see that some electronics firms are now capable to explore the local knowledge sources within the informal *Guanxi* network. However, their capacity to fully transform the informal social asset into higher output and performance is not yet mature. In this aspect, governments can support firms to realize more profit related to high innovation performance by the means of providing innovation funds to resource-limited SMEs and regulating the domestic market that stabilize the reciprocal exchange among the firms.

Theoretical literature has discussed a lot on the issue of proximity and its relationship with learning behaviors, but the empirical support is not yet sufficient to support its role in innovation in different contexts, especially that in developing countries. This chapter takes the step in measuring the use of two most relevant proximities—organizational proximity and social proximity—in the context of China, and adopts a comprehensive view by relating the external interactive learning behavior to the internal absorptive capacity. By responding to the call of bridging the spatial scales of knowledge transfer and learning (Bunnell and Coe 2001; Asheim and Isaksen 2002; Freeman 2002; Fromhold-Eisebith 2007), the chapter has thrown light on the role of proximity at both global and local scale in attaining trust and understanding in the process of interactive learning

However, the complementary role of organizational proximity with global partners and social proximity with local partners is not simple. Actually, as demonstrated by Humphrey and Schmitz (2002), different degrees of organizational proximity, i.e. different ways of insertion into the global production system, influence the local upgrading strategies. Therefore, qualitative studies, e.g. the company and expert interviews, should be conducted to give further insight into the strategic combination of different proximities to achieve the optimal innovation outcome. Moreover, components of absorptive capacity should be further investigated to better understand its relationship with external learning activities. In particular, the exploratory nature of this study points to the need for a more refined design of the R&D indicator in the Chinese context, in which its content and implication is different from that in developed countries.

References

Aghion P, Howitt P (1998) Endogenous growth theory. MIT Press, Cambridge
Allen TJ (1977) Managing the flow of technology. MIT Press, Cambridge
Armour HO, Teece DJ (1980) Vertical integration and technological innovation. Rev Econ Stat 62(3):470–474
Arora A, Gambardella A (1990) Complementarity and external linkages: the strategies of the large firms in biotechnology. J Ind Econ 38(4):361–379
Arrow KJ (1969) Classificatory notes on the production and transmission of technological knowledge. Am Econ Rev 59(2):29–35
Asheim BT, Isaksen A (2002) Regional innovation systems: the integration of local 'sticky' and global 'ubiquitous' knowledge. J Technol Transf 27(1):77–86
Asheim B, Coenen L, Vang J (2007) Face-to-face, buzz, and knowledge bases: sociospatial implications for learning, innovation, and innovation policy. Environ Plan C Gov Policy 25(5):655–670
Barney JB (1991) Firm resources and sustained competitive advantage. J Manage 17(1):99–120
Barney JB (2001) Resource-based theories of competitive advantage: A ten-year retrospective on the resource-based view. J Manage 27(6):643–650
Bathelt H, Malmberg A, Maskell P (2004) Clusters and knowledge: local buzz, global pipelines and the process of knowledge creation. Progress Human Geogr 28(1):31–56
Becker G (1964) Human capital. The University of Chicago Press, Chicago
Bellandi M, Tommaso MR (2005) The case of specialized towns in Guangdong, China. Eur Plan Stud 13(5):707–729
Benhabib J, Spiegel MM (1994) The role of human capital in economic development evidence from aggregate cross-country data. J Monet Econ 34(2):143–173
Bianchi G (1998) Requiem for the Third Italy? Rise and fall of a too successful concept. Entrep Reg Dev 10(2):93–116
Boschma R (2004) Competitiveness of regions from an evolutionary perspective. Reg Stud 38(9):1001–1014
Boschma R (2005) Proximity and innovation: a critical assessment. Reg Stud 39(1):61–74
Brossard O, Vicente J (2007) Cognitive and relational distance in alliance networks: Evidence on the knowledge value chain in the European ICT sector. Paper presented at the summer DRUID Conference, CBS, Copenhagen, 18–20 June 2007
Bunnell TG, Coe NM (2001) Spaces and scales of innovation. Progress Hum Geogr 25(4):569–589
Cameron G (1996) Catchup and Leapfrog between the USA and Japan. Dissertation, University of Oxford
Capello R (1999) Spatial transfer of knowledge in high technology milieux: Learning versus collective learning processes. Reg Stud 33(4):353–365
Carrincazeaux C, Lung Y, Vicente J (2008) The scientific trajectory of the French School of Proximity: interaction- and institution-based approaches to regional innovation systems. Eur Plan Stud 16(5):618–629
Carter AP (1989) Know-how trading as economic exchange. Res Policy 18(3):155–163
CASTED (Chinese Academy of Science and Technology for Development) (1999–2009, annual) Quanguo Keji Jingfei Touru Tongji Gongbao (Statistical Bulletin of China Technology Expenses). http://www.sts.org.cn/tjbg/tjgb/tindex.asp. Accessed 16 Sept 2014
Clark GL, Feldman MP, Gertler MS (eds) (2003) The Oxford handbook of economic geography. Oxford University Press, Oxford
Coase RH (1937) The nature of the firm. Economica 4(16):386–405
Coe DT, Helpman E (1995) International R & D spillovers. Eur Econ Rev 39(5):859–887
Coe NM, Hess M, Yeung HWC et al (2004) Globalizing regional development: a global production networks perspective. Trans Inst Br Geogr 29(4):468–484
Cohen WM, Levinthal DA (1990) Absorptive capacity: a new perspective on learning and innovation. Adm Sci Q 35(1):128–152

Cohen WM, Levinthal DA (1994) Fortune favors the prepared firm. Manage Sci 40(2):227–251
Cooke P, Gomez Uranga M, Etxebarria G (1997) Regional innovation systems: institutional and organisational dimensions. Res Policy 26(4–5):475–491
CSSB (China State Statistical Bureau) (2009) Zhongguo Tongji Nianjian (China Statistical Yearbook). China Statistics Press, Beijing
Delmar F, Davidsson P, Gartner W (2003) Arriving at the high growth firm. J Bus Ventur 18(2):189–216
Doloreux D, Parto S (2005) Regional innovation systems: current discourse and challenges for future research. Technol Soc 27(2):133–153
Eaton J, Gutierrez E, Kortum S (1998) European technology policy. Econ Policy 13(27):403–438
Ernst D (2002) The new mobility of knowledge: digital information systems and global flagship networks. In: Latham R, Sassen S (eds) Cooperation and conflict in a connected world. Routledge, London
Ernst D, Kim L (2002) Global production networks, knowledge diffusion, and local capability formation. Res Policy 31:1417–1429
Evenson RE, Westphal LE (1995) Technological change and technology strategy. Handb Dev Econ 3(1):2209–2299
Feinberg SE, Gupta AK (2004) Knowledge spillovers and the assignment of R & D responsibilities to foreign subsidiaries. Strat Manage J 25(8–9):823–845
Forbes N, Wield D (2002) From followers to leaders: managing technology and innovation in newly industrializing countries. Routledge, New York
Freeman C (2002) Continental, national and sub-national innovation systems—complementarity and economic growth. Res Policy 31(2):191–211
Fromhold-Eisebith M (2007) Bridging scales in innovation policies: how to link regional, national and international innovation systems. Eur Plan Stud 15(2):217–233
Gambardella A (1992) Competitive advantages from in-house scientific research: the US pharmaceutical industry in the 1980s. Res Policy 21(5):391–407
Gereffi G, Humphrey J, Sturgeon T (2005) The governance of global value chains. Rev Int Polit Econ 12(1):78–104
Glasmeier A (1991) Technological discontinuities and flexible production networks: the case of Switzerland and the world watch industry. Res Policy 20(5):469–485
Goedhuys M (2007) Learning, product innovation and firm heterogeneity in developing countries: evidence from Tanzania. Ind and Corp Chang 16(2):269–292
GPBS (Guangdong Provincial Bureau of Statistics) (2009) Guangdong Tongji Nianjian (Guangdong Statistical Yearbook). China Statistics Press, Beijing
Granovetter M (1985) Economic action and social structure: the problem of embeddedness. Am J Sociol 91(3):481–510
Griffith R, Redding S, Van Reenen J (2003) R & D and absorptive capacity: theory and empirical evidence. Scand J Econ 105(1):99–118
Griffith R, Redding S, Van Reenen J (2004) Mapping the two faces of R & D: productivity growth in a panel of OECD industries. Rev Econ Stat 86(4):883–895
Grossman GM, Helpman E (1990) Trade, innovation, and growth. Am Econ Rev 80(2):86–91
Grossman GM, Helpman E (2002) Integration versus outsourcing in industry equilibrium. Q J Econ 117(1):85–120
Hadjimichalis C (2006) The end of third Italy as we knew it? Antipode 38(1):82–106
Harrigan J (1999) Estimation of cross-country differences in industry production functions. J Int Econ 47(2):267–293
Hennart JF (1993) Explaining the swollen middle: why most transactions are a mix of market and hierarchy. Organ Sci 4(4):529–547
Howells JRL (2002) Tacit knowledge, innovation and economic geography. Urban Stud 39(5–6):871–884
Howitt P (2000) Endogenous growth and cross-country income differences. Am Econ Rev 90(4):829–846

References

Humphrey J (2004) Local upgrading in global chains: recent findings. Paper presented at the DRUID Summer Conference, Elsinore, 14–16 June 2004

Humphrey J, Schmitz H (2002) How does insertion in global value chains affect upgrading in industrial clusters? Reg Stud 36(9):1017–1027

Iammarino S, McCann P (2006) The structure and evolution of industrial clusters: transactions, technology and knowledge spillovers. Res Policy 35(7):1018–1036

Ivarsson I, Alvstam CG (2005) The effect of spatial proximity on technology transfer from TNCs to local suppliers in developing countries: the case of AB Volvo in Asia and Latin America. Econ Geogr 81(1):83–111

Jaffe AB (1986) Technological opportunity and spillovers of R & D: evidence from firms' patents, profits, and market value. Am Econ Rev 76(5):984–1001

Keller W (1996) Absorptive capacity: on the creation and acquisition of technology in development. J Dev Econ 49(1):199–227

Kindleberger C (1964) Economic growth in France and Britain, 1851–1950. Harvard University Press, Cambridge

Kirat T, Lung Y (1999) Innovation and proximity—territories as loci of collective learning processes. Eur Urban Reg Stud 6(1):27–38

Lall S (1992) Technological capabilities and industrialization. World Dev 20(2):165–186

Laursen K, Foss NJ (2003) New human resource management practices, complementarities and the impact on innovation outcome. Camb J Econ 27: 243–263

Lazaarini SG, Chaddad FR, Cook ML (2001) Integrating supply chain and network analyses: the study of netchains. J Chain Netw 1(1):7–22

Leibenstein H (1968) Entrepreneurship and development. Am Econ Rev 58(2):72–83

Long JS (1997) Regression models for categorical and limited dependent variables. Sage, Thousand Oaks

Lovett S, Simmons LC, Kali R (1999) Guanxi versus the market: ethics and efficiency. J Int Bus Stud 30(2):231–247

Lucas RE Jr (1990) Supply-side economics: an analytical review. Oxf Econ Pap 42(2):293–316

Lundvall BA (2005) Interactive learning, social capital and economic performance. Paper presented at the international conference on advancing knowledge and the knowledge economy, national academies, Washington, DC, 10–11 January 2005

Lundvall BA, Johnson B (1994) The learning economy. Ind Innov 1(2):23–42

Luo Y (1997) Guanxi and performance of foreign-invested enterprises in China: an empirical inquiry. Manage Int Rev 37(1):51–70

Luthje B (2004) Global Production networks and industrial upgrading in China: The case in electronics contract manufacturing. Paper presented at the international conference on Multinationals in China—Competition and Cooperation, Sun Yat-sen University, Guangzhou, China, 9–11 July 2004

Malmberg A (1997) Industrial geography: location and learning. Progress Hum Geogr 21(44):573–582

Malmberg A, Maskell P (2006) Localized learning revisited. Growth Chang 37(1):1–18

Maskell P (1998) Low-tech competitive advantages and the role of proximity. Euro Urban Reg Stud 5(2):99–118

Massard N, Mehier C (2009) Proximity and innovation through an 'Accessibility to Knowledge' lens. Reg Stud 43(1):77–88

Menzel MP (2006) Dynamic Proximities. Towards a concept of changing relations. Papers presented at the 5th Proximity Congress, Bordeaux, 28–30 June 2006

Meyer S, Schiller D, Revilla Diez J (2009) The Janus-Faced economy: Hong Kong firms as intermediaries between global customers and local producers in the electronics industry. Tijdschr Voor Econ En Soc Geogr 100(2):224–235

Michaely M, Papageorgiou D, Choksi A (eds) (1991) Liberalizing foreign trade: lessons of experience in the developing world. Blackwell, Cambridge

Michie J, Sheehan M (1999) HRM practices, R & D expenditure and innovative investment: evidence from the UK's 1990 workplace industrial relations survey (WIRS). Ind Corp Chang 8(2):211–234

Morrison A, Pietrobelli C, Rabellott R (2008) Global value chains and technological capabilities: a framework to study learning and innovation in developing countries. Oxf Dev Stud 36(1):39–58

Nelson RR, Winter SG (1982) An evolutionary theory of economic change. Harvard University Press, Bonston

Nelson RR, Denison E, Sato K et al (1966) Investment in humans, technological diffusion, and economic growth. Am Econ Rev 56(2):69–82

Nieto M, Quevedo P (2005) Absorptive capacity, technological opportunity, knowledge spillovers, and innovative effort. Technovation 25(10):1141–1157

North DC (1990) A transaction cost theory of politics. J Theor Polit 2(4):355–367

Organization for Economic Co-operation and Development (OECD) (1998) Main science and technology indicators. OECD, Paris

Organization for Economic Co-operation and Development (OECD) (2005) Oslo manual: Guidelines for collecting and interpreting innovation data, 3rd edn. OCED, Paris

Pan YG, Chi PSK (1999) Financial performance and survival of multinational corporations in China. Strat Manage J 20(4):359–374

Park SH, Luo YD (2001) Guanxi and organizational dynamics: organizational networking in Chinese firms. Strat Manag J 22(5):455–477

Peng MW, Wang DYL, Jiang Y (2008) An institution-based view of international business strategy: a focus on emerging economies. J Int Bus Stud 39(5):920–936

Porter ME (2000) Location, clusters, and company strategy. Oxford University Press, Oxford

Piore MJ, Sable FC (1984) The second industrial divide: possibilities for prosperity. Basic books, New York

Ramasamy B, Goh KW, Yeung MCH (2006) Is Guanxi (relationship) a bridge to knowledge transfer? J Bus Res 59(1):130–139

Romer PM (1986) Increasing returns and long-run growth. J Polit Econ 94(5):1002–1037

Romijn H, Albaladejo M (2002) Determinants of innovation capability in small electronics and software firms in southeast England. Res Policy 31(7):1053–1067

Rosenberg N (1982) Inside the black box: technology and economics. Cambridge University Press, Cambridge

Saxenian A, Hsu JY (2001) The Silicon Valley–Hsinchu connection: technical communities and industrial upgrading. Ind Corpo Chang 10(4):893–920

Scherer FM (1998) The size distribution of profits from innovation. Econ Econom Innov 49/50: 495–516

Schmitz H (1995) Small shoemakers and fordist giants: Tale of a supercluster. World Devel 23(1):9–28

Schmitz H, Nadvi K (1999) Clustering and industrialization: introduction. World Dev 27(9):1503–1514

Schumpeter JA (1942) Capitalism, socialism and democracy. Harper & Row, New York

Shein EH (2004) Organisational culture and leadership. Jossey-Bass, San Francisco

Smith JK, Schnucker C (1994) An empirical examination of organizational structure: the economics of the factoring decision. J Corp Financ 1(1):119–138

Standifird SS, Marshall RS (2000) The transaction cost advantage of guanxi-based business practices. J World Bus 35(1):21–42

Storper M (1995) The resurgence of regional economies, ten years later: the region as a nexus of untraded interdependencies. Eur Urban Reg Stud 2(3):191–221

Storper M, Venables AJ (2004) Buzz: face-to-face contact and the urban economy. J Econ Geogr 4(4):351–370

Su CT, Littlefield JE (2001) Entering guanxi: a business ethical dilemma in Mainland China? J Bus Ethics 33(3):199–210

Teece DJ (1986) Transactions cost economics and the multinational-enterprise—an assessment. J Econ Behav Organ 7(1):21–45

Torre A, Gilly JP (2000) On the analytical dimension of proximity dynamics. Reg Stud 34(2):169–180

References

Torre A, Rallett A (2005) Proximity and localization. Reg Stud 39(1):47–59

Tripsas M (1997) Unraveling the process of creative destruction: complementary assets and incumbent survival in the typesetter industry. Strat Manage J 18(summer special issue):119–142

Tsang MC, Rumberger RW, Levin HM (1991) The impact of surplus schooling on worker productivity. Ind Relat A J Econ Soc 30(2):209–228

Vicente J, Suire R (2007) Informational cascades versus network externalities in locational choice: evidence of 'ICT clusters' formation and stability. Reg Stud 41(2):173–184

Vinding AL (2006) Absorptive capacity and innovative performance: a human capital approach. Econ Innov New Technol 15(4&5):507–517

Wang JY, Blomstrom M (1992) Foreign investment and technology transfer: a simple model. Eur Econ Rev 36(1):137–155

Williamson OE (1975) Markets and hierarchies: analysis and antitrust implications: a study in the economics of internal organization. Free Press, New York

Whitford J, Potter C (2007) Regional economies, open networks and the spatial fragmentation of production. Socio-Econ Rev 5(3):497–526

Yang C (2009) Strategic coupling of regional development in global production networks: redistribution of Taiwanese personal computer investment from the Pearl River Delta to the Yangtze River Delta, China. Reg Stud 43(3):385–407

Yeung HWC (2009) Regional development and the competitive dynamics of global production networks: an East Asian perspective. Reg Stud 43(3):325–351

Zahra SA, George G (2002) Absorptive capacity: a review, reconceptualization, and extension. Acad Manage Rev 27(2):185–203

Zhang Y, Zhang ZG (2006) Guanxi and organizational dynamics in China: a link between individual and organizational levels. J Bus Ethics 67(4):375–392

Zhou X, Li Q, Zhao W et al (2003) Embeddedness and contractual relationships in China's transition economy. Am Sociol Rev 68(1):75–102

Chapter 5
From Globalized Production Systems to Regional Innovation Systems: Governance and Innovation in Shenzhen and Dongguan, China

Abstract Governance constitutes elementary supportive infrastructure for regional innovation system, as they are able to facilitate interaction and cooperation. This chapter extends the evolutionary lens of governance to the transforming process of a regional production system into a regional innovation system, and shows how the formation of regional innovation system has unfolded under two distinct governance modalities. Drawing on the inter-city comparative case from Shenzhen and Dongguan, China, the chapter shows that Shenzhen, characterized as the dirigiste globalized production system in the beginning of opening-up policy, has evolved to a higher level of interactive regional innovation system. On the other hand, the typical city-region for grassroots globalized production system—Dongguan, encounter difficulties in its transformation into a regional innovation system as firm innovation is still rigorously controlled by global lead firms through hierarchical production chains. Finally, the future path for regional upgrading with divergent governance modalities is discussed, particularly for grassroots governance, incentive frameworks should be put in place to avoid the negative lock-in effects associated with local vested interests.

5.1 Introduction

The concept of regional innovation system, which developed from the national innovation system literature, takes institutional and organizational dimension in the territorial level into consideration of innovation activities (Cooke et al. 1997; Howells 1999; Cooke 2001; Revilla Diez 2000; Morgan 2004; Asheim and Coenen 2005). In the analytical framework of a regional innovation system, the institutions and organization are extended as the governance infrastructure that facilitates cooperation, organizes interaction, reduces uncertainty and cuts transaction cost, enabling the business sector to compete more competitively (Cooke 1992; Cooke et al. 1998).

This chapter has been reorganized and revised in a substantial manner and published as a second edition in *Regional Innovation Systems within a transitional context: Evolutionary comparison of the electronics industry in Shenzhen and Dongguan since the opening of China*. Fu W, Revilla Diez J, Schiller D. *Journal of Economic Surveys 26/3*, Copyright © 2012, Wiley-Blackwell.

Cooke (1992) proposes three modalities of governance supporting the business inter-relationships: grassroots governance, network governance and dirigiste governance. These three modalities of governance differ in the degree of policy intervention as well as the relationship with knowledge-intensive organizations at different scales. Cooke et al. (2004) revisit the regional innovation system first proposed in a systemic way in late 1990s (Braczyk et al. 1998) with an evolutionary perspective in the face of monumental economic shift and uncertainty. In the practice of many regional innovation systems around the world, the governance infrastructure evolves according to the needs of market change and industrial organizational restructuring, aiming at generating more dynamic regional growth mechanisms.

When this line of thinking on evolving governance infrastructure extends to the context of latecomer countries, where the regional innovation system is itself burgeoning from the production system relying heavily on integration into the lower-end of global production networks, the evolutionary lens should be expanded beyond the scope of regional innovation systems. That is to say, the focus on the evolution of governance infrastructure should be put on the transition from governance that supports initial industrialization to governance that supports the innovation activities.

This chapter aims to understand how different governance infrastructure influence upon the development of regional innovation systems by investigating two cities in Southern China where initial industrialization has been supported with different modalities of governance following the introduction of the opening policy. In Shenzhen, the governance supporting industrialization is rather dirigiste, characterized by a state-oriented involvement of economic development with ex-ante strategic policy support. In Dongugan, however, governance that supports industrialization is grassroots, characterized by flexible institutions organized mainly by town and village authorities that are favorable for overseas Chinese investment based on *Guanxi* (Leung 1993; Yang 2012).

Thanks to the state initiative to develop electronics production at the very beginning of the establishment of special economic zone, the electronics industry gained a first mover advantage in Shenzhen compared to Dongguan, despite the fact that both faced opportunities for a global shift of the processing function to low-cost areas. Dongguan then followed up when the electronics industry replaced the old primary textile industries in the 1990s. With the rising land and labor prices as well as the fierce competition from other low-cost areas, policy reaction was initiated at various levels of government, aiming to form a network governance to support the upgrading and innovation activities of the firms and regions.

However, the empirical analysis of self-conducted electronics firm survey in 2009 reveals different business innovation pattern in Shenzhen and Dongguan. In Shenzhen, the regional innovation system displays an interactive feature. Firms are capable of interacting with a wide range of external partners to promote innovation outcomes. In contrast, the regional innovation system in Dongguan is heavily dependent on global lead firms. The scope of interaction and learning related to innovation among Dongguan firms is limited to tacit knowledge from organizationally proximate parent companies and foreign customers.

This different pattern of business innovation can be explained with two aspects of governance infrastructure from an evolutionary perspective: endowments of innovation supported resources and the negative lock-in effect induced by competency trap and vested interest. In other words, the successful transition from industrialization-led governance to innovation-supported governance depends on the competence in accumulating and mobilizing the innovation-related resources.

The chapter is organized as follows. Section 5.2 elucidates governance infrastructure in production and innovation systems. Moreover, theoretical discussion from an evolutionary perspective will be provided on what facilitates or handicaps the evolution of governance infrastructure for low-end production to support of innovation. Section 5.3 presents the survey design of the comparative investigation into the feature and level of the regional innovation system under different governance modalities. Section 5.4 depicts the governance modalities in Shenzhen and Dongguan in the initial industrialization phase and the transitional phase. Overall descriptive innovation indicators for Shenzhen and Dongguan are displayed in Sect. 5.5. In Sect. 5.6, empirical results are demonstrated based on questionnaire data from electronics firms in order to explore innovation pattern in Shenzhen and Dongguan. Finally, Sect. 5.7 concludes and discusses the policy implication derived from the cases in Shenzhen and Dongguan.

5.2 Evolutionary Regional Innovation System and Governance Infrastructure

5.2.1 Evolution of Governance Infrastructure: Content and Typology

Governance consists of relations of power and structures of decision-making to coordinate the input-output production system (Storper and Harrison 1991). Reform of governance has been found to be the catalyst of rapid industrialization in latecomer countries (Goldsmith 2007). In latecomer countries, the governance has been adjusted and developed to match the external needs due to the great dependency on external market and technology. Successful operational outcome also depends on the institutional fit between local politics and transnational corporation (Yeung 2000).

Governance aiming at launching and supporting industrialization covers three aspects, as shown by Table 5.1. The governance in production system has no explicit innovation content, in which the focus is mainly on initiating the growth of production and export activities and support it with various measures. The organizations that carry out these tasks in the old production system might refer to government functional offices, industrial associations, folk unions and state-owned large companies. In the Chinese context, these organizations orchestrate under the common aims of boosting local GDP.

Table 5.1 Governance in production and innovation system. (Cooke et al. 1997)

	Production system	Innovation system
Institutional competence	Capacity to design and execute industrial development policies	Capacity to organize technology transfer & to support science and technology program
Supported infrastructure	Hard infrastructure such as roads, electricity, port, etc.	Density and quality of infrastructures for innovation such as universities, research institutes, technology transfer agencies, consultants and skill-development and training agencies
	Soft infrastructure such as administrative services to assist the firms	Control or shared execution of part of strategic infrastructures
Financing & Budget	Capacity to impose taxes	Accessibility for firms to capital market
	Autonomy for public spending	High level of financial intermediaries

When the spatially specialized entity evolves into an innovation system, the governance should co-evolve and adjust the focus to supporting innovation activities. To secure systematic learning and innovation synergies that occurs externally of the firm boundary, governance plays an important role in providing access to information, ensuring credibility, coordinating collective actions and even creating a learning atmosphere (Dalum et al. 1992; Sweeney 1991; Amin 1999; Haggard 2004).

In accordance with the governance elements in production system, Cooke et al. (1997)' outline the governance dimension in regional innovation systems as follows: (1) Institutional competence to organize technology transfer and launch science and technology programs; (2) Supported infrastructure to enhance the capacity of innovation and extend the scope of interactive learning; (3) Financial and budgetary capacity to reduce innovation-related uncertainty and risk as well as mobilize innovation-related resources.

Institutional Competence In a globalizing economy, institutional setup should co-evolve and echo with the external industrial trend at national and global level such as vertical disintegration and division of labor. Due to the great dependency of latecomer countries on external market, the institutional setup has been adjusted and developed in response to match the external needs. Yeung (2000) elucidates that "institutional fit" between local politics and transnational corporation should be achieved to successful operational outcome. Therefore, the external market and trend of industrial organization should be born in mind when institutional competence is taken into consideration.

Institution is also embodied as the capacity of innovation policy to strategically identify new and related industries that might trigger and extend the scope of interactive learning and systematic innovation. The trigger effect of new and related industries is verified by the Jacobs (1969) and Boschma and Iammarino's (2009) relatedness of knowledge. Demonstrated by the research over regional innovation system in Europe and America, strong innovative performance is mostly accompanied by strongly developed territorial administration which involves the intervention of public organizations (Cooke 2001). This aspect of institution capacity is of great relevance in rapidly industrialization context in China where new industries are introduced and induced by government policies aiming at attracting FDI.

Supported Infrastructure Firms and related institutes are important components of the innovation capacity in the system of innovation. Although in regional innovation system, the scope of institutions goes beyond the knowledge-intensive ones and starts to pay attention to all institutions that define the way innovation actors interact, the density and quality of knowledge-intensive institutions are important for systematic relations to take shape. As innovation-induced interaction and learning requires the complementary knowledge and sufficient internal absorptive capacity among the various actors, institutions—especially the knowledge-related ones such as universities, research institutes, technology transfer agencies, consultants and skill-development agencies—expand the scope of complementary knowledge and interactive learning that the innovative firms are able to draw upon to foster innovation (Asheim and Isaksen 2002; Asheim and Coenen 2005).

Among these innovation infrastructures, innovative firms are the core element. It is actually the willingness and capacity to interact with external partners that determines the degree of systematic innovation. From an evolutionary view, the knowledge base of a place influences the entrepreneurial capacity and human capital that are all closely related to the innovativeness of firms in the developmental path.

Financial and Budgetary Capacity Innovation activities bear uncertainty and risk on the return of huge amount of capital investment on equipment, training, marketing, etc. Firms, especially small and medium ones, are able to conduct innovation more thoroughly when it is easy and efficient for them to resort to external financing.

In territorial level, the autonomy of financial and budgetary capacity is elementary for the incentive to develop innovation infrastructure and institutional competence. Other than direct financing such as funds and loans, institution in the limited territorial scale can be used to minimize the uncertainties between the lenders and borrowers by informal social orders or formal regulations. In this aspect, information sharing is important in successful financing activities (Cooke et al. 1997).

In terms of governance content, three typologies of regional innovation systems (RIS) can be drawn according to Braczyk et al. (1998): the grassroots, network and dirigiste governance modalities.

Grassroots RIS In terms of institutional competence in this modality, the initiation process of technology transfer and technology programs are organized at the town or district level, and degree of supra-local co-ordination is low because of the localized nature of organization. In terms of supported infrastructure, the research competence is highly applied or near-market. Moreover, the level of technical specialization will be low, lacking finely-honed expertise. Funding in grassroots RIS comprises a mix of capital, grants and loans from local banks, local government and possibly local Chamber of Commerce.

Network RIS In terms of institutional competence in this modality, initiation process of technology transfer and technology programs are organized in multi-levels, encompassing local, regional, federal and supranational levels. Also, system coordination is high because of the large number of stakeholders and the presence of associations, fora, industry clubs and the like. In terms of supported infrastructure, the research competence is a mix of both pure and applied knowledge geared to the needs of large and small firms. Funding in network RIS is guided by agreement among banks, government agencies and firms at national, regional and local levels.

Dirigiste RIS In terms of institutional competence in this modality, initiation process of technology transfer and technology programs is a product of central government policies, and the degree of coordination is high since it is state-run. In terms of supported infrastructure, research is rather basic or fundamental and relates more to the needs of larger (possibly state-owned) firms. Funding in dirigiste RIS is largely centrally determined although the agencies may have decentralized by setting up local departments in the regions.

The innovation governance supports the firms with diverse posture in the market place with producers and customers, ranging from a global to local reach (Braczyk et al. 1998). Firms could organize production and innovation in accordance with the governance support in a localized, interactive and globalized manner. The evolutionary investigation of most case regions in Cooke et al. (2004) indicates a trend towards interactive business innovation, which responds to the emphasis on interactive learning and systematic innovation in modern innovation theories (Cooke et al. 1997; Howells 1999; Revilla Diez 2000; Asheim and Coenen 2005; Lundvall 1992). In interactive business innovation, a high degree of association has been formed vertically and horizontally with both global and local reach. In this way, innovative synergy forms gradually, hatching the knowledge spillover and spatial dynamic externalities among firms.

This chapter focuses the analysis of two cities in coastal Southern China, where the production system in the initial industrialization phase is dominated by global corporations with clustered supply chain of dependent SMEs. What is more, since the advent of the opening policy in the late 1970s, the central government has either been directly involved in economic development, such as establishing special economic zones, or has implicitly encouraged the bottom-up development, mainly by transferring more economic developmental autonomy to local governments. Therefore, the theoretical discussion in the following Sect. 5.2.2 centers around the dynamics and inertia faced by the evolution of grassroots globalized production system and dirigiste globalized production system towards a well-functioning innovation system. In Fig. 5.1, the evolutionary paths of the grassroots globalized production system in Dongguan towards an innovation system, as well as that of the dirigiste globalized production system in Shenzhen, are illustrated. The underlying logic in this figure is taken from an evolutionary perspective, which argues that the initial governance modality in the initial phase is able to exert significant influence on the transformation from a simple-production-supported territorial system to something of a more innovation and entrepreneurship-supported one. The following theoretical analysis and empirical investigation would justify this argument.

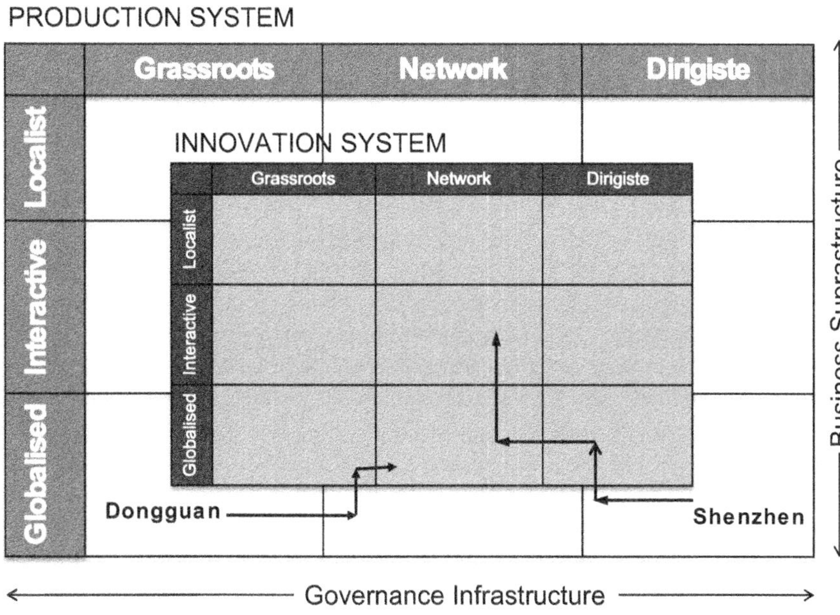

5.2 Evolutionary Regional Innovation System and Governance Infrastructure

PRODUCTION SYSTEM

Fig. 5.1 Evolution from production system to innovation system

Note: Arrow indicates direction of system movement 1978-2009

5.2.2 Evolution of Governance Infrastructure: Dynamics and Inertia

Governance can refer to two interrelated aspects: institutions and organizations. Institutions are the rules of the game and organizations are embedded in the institutions, playing the game with different competences and capabilities (Cooke et al. 1998). The interaction between institutions and organizations defines the evolutionary path of governance infrastructure. In other words, the institution defines the behavior of organizations, and organizations have a return influence upon institutions by adjusting them to meet the needs of the changing external environment.

The dynamics of the governance evolution towards innovation-supported depends on the capability of the organizations (Martin 1999). The critical mass, referring to the networked organizations with given capabilities and resources, together shapes and strengthens the governance structure for economic activities (Stam and Lambooy 2012). In the initial industrialization phase, when the industrial base is weak, the perspective of resource endowments of related organizations becomes an important baseline for the evolution of governance towards a well-functioning innovation system. In Porter's (1998) competitive model, local endowments such as highly specialized skills and knowledge, institutions, related businesses and demanding customers are emphasized for the construction of a competitive cluster.

For a grassroots globalized production system, production capital and know-how depends heavily on foreign investment. There is very limited skill base in the production system either from previous accumulation or assignments from the central, which would hinders the absorption of foreign technology transfer and spillover. In contrast, the dirigiste globalized production system is able to accumulate the skill and knowledge stock from the central assignments such as relocation of large state-owned firms and research institutes. Levels of foreign investment in the local environment thus differs under these two different governance modalities, which defines the capacity of localities to process, absorb and exploit the external information and technological spillover in the export-oriented regions (Cohen and Levinthal 1990; Gambardella 1992; Tripsas 1997; Zahra and George 2002). In the joint ventures between foreign investors and capable state-owned firms, which possess stronger stock of knowledge and capital compared to the small township and village enterprise in the initial industrialization phase, engineers are able to accumulate modern production experience thanks to the economies of scale and scope. And these employees constitute the potential pool of human capital and entrepreneurship for future development in the private sectors (Kim 1999).

Furthermore, the capability of dirigiste governance to bring new dynamics into the economy is well reflected by the technology foresight argumentation. According to the practices in some countries such as Japan, Britain, Australia and New Zealand (Martin and Johnston 1999), technology foresight, which is mostly conducted by government agencies or advisory boards, generates concentration on long-term development of selected trajectories and develops a level of consensus on desirable futures. The successful economic growth in Korea since the mid-1960s well illustrated the strength of dirigiste approach, under which heavy and chemical industry was strategically planned and took on rapid development in export market (Eshag 1991). Technology foresight includes the practice of selecting technology priorities, identifying new strategic industries, creating partnership between sciences, industry and government, as well as providing incentives for multidisciplinary research. Under some extreme circumstance, crisis construction can be applied to force the firms to undertake challenging tasks (Kim 1999). Therefore, the dirigiste approach, which is mostly initiated and governed by national level agencies with more power, is more able to draw on technology foresight to inject new dynamics into development than the grassroots approach. Especially in the time of rapid technology regeneration, a grasp of future trends and timely reactions are important for the region to keep a dynamic growth path.

Although the dirigiste globalized production system possesses more knowledge and skill endowment and is more able to draw on technological foresight than the grassroots approach, it is still insecure to leave the future of development in the hands of central authorities. Firstly, there might be misinvestment in the selection of key industries when little information is collected from the market, generating negative opportunity costs for the locality. Secondly, soft budget-constraints are mostly likely to occur in state-owned firms, which play an important role in the dirigiste approach, causing lower efficiency and poorer performance than in private sectors (Qian and Roland 1998). Therefore, there is an urgent need for dirigiste modalities to evolve towards network governance, involving more market mechanism of competition. In addition, the participation of market power would incentivize the exploitation of entrepreneurial activities on the accumulated knowledge stock in dirigiste production systems, enabling the firms to undertake interactive learning to gain innovation ideas and related support.

On the other hand, the evolution of grassroots governance from a production system to an innovation system carries more inertia than the dirigiste one. As argued by Easterly (2008), the grassroots approach evolves gradually within the constraint of previous institutions, while the dirigiste approach is able to start with a blank sheet or even tear up the old institutional setup. This argument has two implications. Firstly, while the dirigiste governance is able to draw on technology foresight, a "competency trap" might arise in grassroots governance, as being too good at something constrains the capacity of grassroots organizations to absorb new ideas and develop new trajectories (Levitt and March 1988). In the light of this, a mixed level of organizations should be in place to ensure breaking through the "sticky knowledge" and forming new competencies. Secondly, vested interests in organizations tend to emerge in the evolving process of the grassroots governance, which might oppose the changes that undermines their current gains and positions (Boschma 2004). This aspect is demonstrated by the restructuring problems that are faced by previously heavily industrialized areas in Britain and Germany (Hudson 1994). Altogether, it constitutes the "cognitively sunk cost", which creates a negative reinforcing cycle, impeding new development dynamics and trajectories (Leonard-Barton 1992).

Therefore, grassroots governance in a production system with a weak industrial base tends to encounter competency traps and complex vested interests, leading to the risk of negative lock-in and sticky inertia. When governance evolution towards the one supporting innovation systems encounters inertia in the face of restructuring and upgrading, it would create systemic market and policy barriers to interactive business innovation as new development alternatives (Könnölä et al. 2006).

The theoretical overview of the governance infrastructure discussed above provides the starting point for investigating its impact on business innovation activities. The comparison of the Shenzhen and Dongguan cases should reveal a different pattern of interactive learning and systemic innovation, providing a divergent evolving path of governance infrastructure as shown by Fig. 5.1. Before addressing the innovation pattern based on empirical results, the research design will be presented, followed by the review of the evolving governance infrastructure in Shenzhen and Dongguan since the opening policy in 1978 and the current overall innovative performance in the two cities.

5.3 Survey Design of a Comparative Study

The comparative study has been proposed by many scholars, for example Staber (2001), Doloreux (2002), Dolereux (2004) and Asheim and Coenen (2005), as the most important means of fully understanding the function of regional innovation

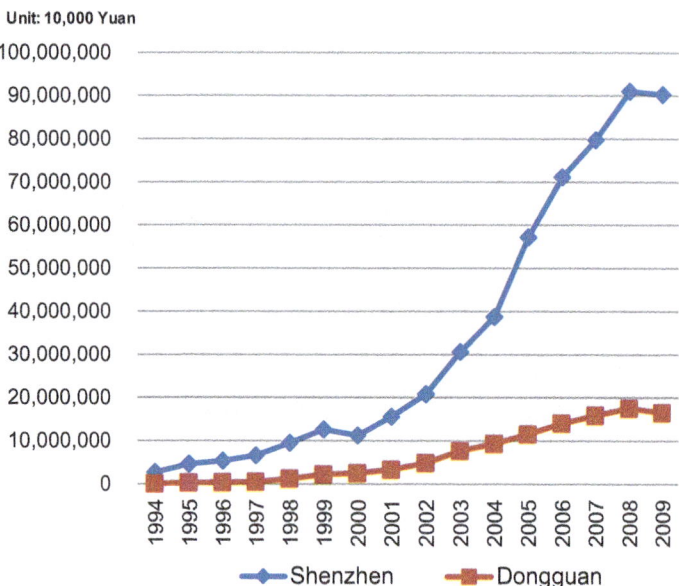

Fig. 5.2 Development of electronics industry in Shenzhen and Dongguan. (Data source: SMBS (1995–2010, annual); DMBS (1995–2010, annual))

system and capturing hidden variables that are of interest to its construction. Therefore, comparing the evolution of the regional innovation systems in Shenzhen and Dongguan offers a unique perspective for understanding the specific contents of governance infrastructure that influence the systemic innovation in the region.

The empirical data were collected from an electronics firm questionnaire survey in Shenzhen and Dongguan, Guangdong Province, China. The investigation focuses on the electronics industry because of its great dominance and development history in the research area, which enables the inquiry into its evolutionary path. As shown in Fig. 5.2, the output value of the electronics industry in Shenzhen and Dongguan kept growing during the period between 1994 and 2009. Dongguan, which is known as the world factory of electronics product, experienced a much lower level of output value growth than Shenzhen due to the concentration of lower value processing.

The questionnaire survey was conducted via telephone and mail in order to ensure the feasibility of the survey and validity of the data, and was strengthened by following-up that aimed to persuade the firms to fill out and send back the questionnaires, as well as to fill out unanswered questions after the questionnaires were returned. Additionally, in order to establish contact with more firms, a second approach was applied, namely visiting fairs. The fairs and firms were randomly selected. Moreover, the fairs visited have a large number of firm exhibitors, ensuring the unbiased nature of the fair-visiting result. In total, 312 Shenzhen firms and 281 Dongguan firms were contacted. In total, 167 Shenzhen firms and 177 Dongguan firms filled out the questionnaires, with the response rate in Shenzhen and Dongguan being 54 and 63 %, respectively.

Viewing from descriptive statistics of the surveyed sample, there are 140 innovative Shenzhen firms and 161 Dongguan firms. In the questionnaire, the constructed variables regarding the innovation activities of electronics firms cover the internal efforts and external interaction during the innovation process, i.e. acquiring new innovative ideas, acquiring codified knowledge and tacit knowledge. The scope of external interaction covers various business partners, such as parent companies, foreign customers, domestic customers, universities and research institutes, as well as sales agents. In addition, the informality of interaction with the partners is identified, i.e. interacting with the partners through active search strategy such as the Internet, exhibitions or sales agents, or interacting with the partners through the introduction and recommendation of long-term business partners, relatives and friends. Surveyed firms were asked about the importance of each aspect in product innovation activities on a one to five likert scale. The reliability analysis of the total 21 innovation-related items has been run, and the Cronbach's Alpha value has reached 0.85 which suggests the consistency and reliability of the questionnaire design.

5.4 Governance in Shenzhen and Dongguan, China: An Evolutionary Overview

The institutional setups in Shenzhen and Dongguan, which have evolved since the open door policy to meet the needs of rapid industrialization, correspond to the dirigiste and grassroots governance modalities respectively. In the following analysis, the evolution process of governance will be summarized by the thorough review of the "Shenzhen Electronics Yearbook" (SECC 2004, most of the information is systematically summarized in Sect. 7.4 Appendix D) and the "Guangdong Electronics Yearbook" (GECC 2002). In these two yearbooks, descriptive facts are provided for the developmental path of the electronics industry in Shenzhen and Dongguan. Moreover, an in-depth interview was conducted in late 2007 with the former president of Dongguan Electronics Association to gain insight into the industrial development history and changing interests of governments at various levels.

5.4.1 Governance Evolution in Shenzhen Since Opening

5.4.1.1 Shenzhen's Governance in the Initial Phase of Industrialization

Before it was selected by the central government as one of the special economic zones in 1979, Shenzhen was a small, peripheral town. The opening policy has been brought into full play in Shenzhen owing to its superior locational advantage adjacent to Hong Kong. The role of the electronics industry was in focus from the very beginning of the special zone development in Shenzhen (GECC 2002; SECC 2004). Due to its geographical proximity to Hong Kong, the electronics industry has been developing rapidly relying on processing operation.

Governance to initiate industrial development is based on the strategy of merging large-scale foreign investment with large state-owned firms that possess good resources endowments. Favorable policy for attracting foreign investment is designed to encourage large-scale programs with longer fund turnover periods, aiming to control short-term opportunist behavior of foreign firms.

Special financial formulas, such as joint ventures between large state-owned companies and foreign investors, are applied. These large firms were originally an important part of the national innovation system in the planned economy. They stemmed from large state-owned companies directly under the jurisdiction of state ministries and provinces, renowned universities and research institutes, as well as military-related plants that were highly specialized in heavy industry. Right from the opening in 1979, many divisions of ministry- and province-subordinated firms and institutes have been agglomerating in Shenzhen rapidly due to order from central government to develop the special zone (SECC 2004).

These state-owned firms played an important role in organizing and nurturing the industrial cluster in the very beginning of development. On one hand, they were heavily endowed with highly qualified human capital and technology that were leading among Chinese counterparts at that time. On the other hand, they struggled with the low profit due to irrational ownership incentive and were therefore thirsty for external capital and global leading technology. At that time, they were then able to introduce high-scale production lines due to the disposal of state-owned assets and scale economies of production. Moreover, the high endowment of human capital in state-owned companies enables the better absorption of imported technology. After taking advantage of foreign capital and technology, the growth of these domestic firms took a considerable rate (SECC 2004).

Besides joint venture with foreign companies, there were also joint ventures between domestic state-owned firms, mainly between the firms under jurisdiction of state ministry and firms under the jurisdiction of Guangdong province. Peng and Heath (1996) points out further the state-owned companies in transition economy, while applying the conventional acquisition and expansion strategy, settle as well on network-based strategy of growth drawing upon personal trust and informal agreement among managers. Moreover, the alliances among these state-own companies were always accompanied by tasks of developing a specific leading technology, e.g. color kinescope, Liquid Crystal Display (LCD), small-volume exchange equipment and multi-layer printed circuit in 1984 as well as optical fiber in 1997 (SECC 2004). In addition, Shenzhen City Government also sponsored the direct investment in high-tech companies to nurture new growth opportunity and attract high-end foreign investment (SECC 2004).

The inter-firm linkages of production, information and technology have been built with the growth of these Chinese firms allying between each other under the state order as well as with the foreign firms. In 1986, the Shenzhen Electronics Group Company (later known as Saige Group), which unifies 117 companies among all the 178 companies in Shenzhen on voluntary basis, was established under the approval of the Shenzhen City Government. It was then one of the four experimental sites of enterprise group of electronics industry in China. In 1988, the Shenzhen Electronics Group Company arranged the construction of the first specialized electronic parts supply market in China, "Saige Electronics Supply Market", which is a remarkable milestone in organizing the supply chain of the electronics industry in Shenzhen (SECC 2004). Within this organizational arrangement, information and production opportunities are more frequently shared among member companies. The Saige Market later gradually becomes the breeding soil of local entrepreneurship in Shenzhen.

Gradually, network governance has been formed in multi-level organizations, encompassing China Central Ministries, Guangdong Province and the Shenzhen City Government and industrial park authorities in the aspects of initiating technology transfer, facilitating technological absorption of domestic firms and assisting the business sector in training, quality control and customer searching (SECC 2004).

With the support of the dirigiste governance and geographical proximity to Hong Kong, the electronics industry in Shenzhen has been developing rapidly relying on simple processing operation in this period. Nevertheless, the industrial structure in electronics was concentrated in the standard consumer electronics industry (mainly telephone, TV, calculator and radio), which was faced with a saturated market and limited space of technological upgrading (SECC 2004).

5.4.1.2 Shenzhen's Governance in the Transitional Phase

After 1990, the electronics industry in Shenzhen faced the rising factor price and gradually lost the technological leadership in consumer electronics compared to the other regions in China. In order to achieve successful upgrading towards high-tech electronics, the Shenzhen city government has also strategically drawn on the technological foresight in five industries: PC and software, telecommunication, microelectronics, optical-electro-mechanical integration and new materials. Under the guidance of the selected industries, foreign investment was supported around the five industry fields (SECC 2004).

For the purpose of adjusting the institutional competence to initiate the upgrading, the Shenzhen Government implemented two primary measures in terms of financing programs. Firstly, firms were offered the accessibility to capital markets, with the first stock market being opened in Shenzhen in 1992. Further in 2006 and in 2009, the Small and Medium Enterprise Board and ChiNet was launched consecutively, making Shenzhen one of the most important financial and entrepreneurial centers. Secondly, the city government supported the small and medium-sized high-tech private firms with specific funding intermediaries, which has led to the growth of technological leading firms such as Huawei and Zhongxing (SECC 2004).

> In 2002, half of the state-level 909 projects on integrated circuit design have located in Shenzhen and a cluster of integrated circuit design companies already took shape, which covers the operation of encapsulation, testing, plate making, device providing, scribing and thick film integrating. Among these firms, most of them are domestic firms such as Guowei, Huawei, Zhongxing, Aisikewei, etc. By the end of 2002, Intel and STMicroelectronics all followed and established research and design center of integrated circuit in Shenzhen.
> —Extracted from SECC 2004.

The foreign investment in Shenzhen was only experimental at first and does not constitute the pulling motor of development in Shenzhen. However, it did bring new management concepts to Shenzhen aside from the advanced equipment. Owing to Shenzhen's special background as the experimental field for opening policies in China, private firms and privatization reform of state-owned firms were encouraged and supported by various levels of government. In 1993, Shenzhen's National People's Congress adopted the "Stock Limited Corporations Ordinance of Shenzhen Special Zone" and "Limited Liability Company Ordinance of Shenzhen Special Zone" with legislative power of the special zone authorities. Even in small and medium sized state-owned companies, employee stock ownership was gradually allowed. Under this circumstance, the human capital endowment was able to be released from the old national innovation system locked within the state-owned companies, central ministries (Shenzhen Division) and research institutes (Shenzhen Division), which altogether enables the exploitation of vast market opportunities in technology. Entrepreneurs propagated in Shenzhen and the young migrants were eager to explore the huge market opportunities in a time of transition, reform and rapid growth. As a result, many private firms flourished in the 1990s, establishing the base for a wide scope of interactive learning and systemic innovation in the interactive regional innovation system.

5.4.2 Governance Evolution in Dongguan Since Opening

5.4.2.1 Dongguan's Governance in the Initial Phase of Industrialization

With the devolution of partial power of fiscal arrangements and foreign investment policies from the central government to town and village governments, the Dongguan local government has been enthusiastically devoting to economic growth since the opening up policy.

The industrialization process in Dongguan started in the garment and shoe industries during 1980s. Compensation trade, i.e. processing raw materials on clients' demands, assembling parts for the clients and process according to the clients' samples, expanded quickly in many villages and towns. The orders were mostly sourced from Hong Kong owing to social and cultural proximity. At that time, there were about 650,000 Dongguanese settled in Hong Kong. They worked or opened their own factories in Hong Kong and thereby were the mostly reliable communicators of business between Hong Kong and their hometown (Interview in Dongguan, September 2007).

The Dongguan local government paid primary attention to encouraging the Hong Kong-Dongguanese to invest in their hometown. In 1981, the office of outward processing and assembly was established to fulfill this important task. Moreover, the village and town governments also greatly supported the development of compensation trade by offering cheap land, favorable policies and flexible standards. The distribution of the processing earnings is negotiated between the town and village governments and foreign investors, mostly under informal frameworks such as oral agreements (Interview in Dongguan, September 2007). In this way, vested interests were taking shape among foreign firms, township and village governments and peasants who live on the rent of the collectively owned properties.

In the process of industrial development based on grassroots foreign investment, infrastructure supply is directed to industry-specific and hands-on service by the township and village governments, who deployed the fiscal income into construction, such as factory buildings, roads, electricity and telecommunications, to improve the investment environment. This bottom-up industrialization process matched simultaneously with small-scale Hong Kong investment that feared institutional uncertainty, leading to the scattered land use pattern and low agglomeration economy. Nevertheless, the demonstration effect of "successful small Hong Kong bosses" and the formation of vested interests have further strengthened the governance focusing on compensation trade in Dongguan.

5.4.2.2 Dongguan's Governance in the Transitional Phase

By 1995, the profit space of garment industries has been greatly shrinking. Electronics firms, mainly led by Taiwanese firms, along with some of the Shenzhen firms, were gradually relocating to Dongguan in the mid-1990s. The shift, attracted by low-cost factors in Dongguan, was systematically carried out through the clustering of Taiwanese firms with complex supplier linkages. Take Delta Electronics for an example, it has brought 22 small and medium sized upstream and downstream Taiwanese firms when investing in Dongguan. Relying on the networked production bought by Taiwanese firms, the electronics industrial chain is now complete and integrated in Dongguan with a kitting rate more than 95 %. At the beginning of the twentieth century, the compensation trade in electronics in Dongguan reached its peak. However, even before its accelerating phase in the mid-1990s, the policy focus at the provincial level on electronics industry development was still primarily placed on Shenzhen, Guangzhou and Foshan, rather than on Dongguan (GECC 2002).

In order to attract large-scale high-tech investment in the face of industrial upgrading and restructuring, the Dongguan City Government established the first city-level industrial park with high entry standards in 2001. Furthermore, the Dongguan City Government responded to the call from the central and provincial governments to "empty the cage for new birds", i.e. to evacuate the old low-end processing industries and attract new high-tech ones. However, this led to great resistance from the township and village governments. On one hand, the township and village governments and the peasants rely heavily on processing firms for their major income (Yang 2012). Therefore, vested interest has been firmly forged from the bottom up, thus creating the inertia for structural change. On the other hand, the township and village governments not only lack the incentive, but also the experience to undertake far-sighted ex ante developmental arrangements and provide necessary infrastructure support—often in a much higher scale than the one supporting simple

Table 5.2 Domestic sector in Shenzhen and Dongguan (2009). (Data sources: SMBS 2010; DMBS 2010)

Firm above designated size[a]	Shenzhen (%)	Dongguan (%)
Share of domestic firm in total firm number	53	25
Share of domestic firms' output value in total output value	37	16
Share of domestic firms' added value in total added value	47	15

[a] Firms above designated size include all state-owned firms and firm with no less than 5 million sales

production—in order to secure upgrading towards high value-added activities (Interview in Dongguan, September 2007).

> The profit of garment industry has been shrinking after 1995, and the development of electronics industry took pace. At that time, the bosses of medium-sized firms in Taiwan saw the huge profit made by the bosses of small-sized firms investing in Dongguan, and decided to follow in and establish plants here ... However, the industry is without planning at all because Dongguan government, especially the town government, would offer land whenever the foreign firms are willing to invest. I remember that many surrounding towns and cities laughed at us on that, calling it 'there are so many stars in the sky of Dongguan but without a moon'.
> —Interview with Dongguan Electronics Association President Ye in 2007[1]

Due to the weak industrial base before the rapid development, the local skilled labor market and related industrial institutions remained underdeveloped, especially in comparison to the great profit made too quickly via compensation trade. Statistics in the year 2009 show that the domestic sector was much weaker in Dongguan than in Shenzhen (Table 5.2). This less endogenous development path is expected to impact on the development of the regional innovation system in Dongguan.

5.4.3 Summary of Governance in Shenzhen and Dongguan

From the above discussion, it can be concluded that the development of the electronics industry in Shenzhen is strongly supported by ex-ante involvement of state authorities and institutes that simultaneously echoed with the trend of the global industrial shift of the electronics industry to low-cost regions in the 1980s and market institutional reform in the vanguard of China's Special Economic Zones (Luthje 2004). On the other hand, the institutional setup in Dongguan has repeatedly been strengthened for the aim of processing trade development with the symbiotic gain of the village and township level governments, overseas Chinese investors (mainly Hong Kong and Taiwan) and local peasants. Moreover, the support of institutional organizations is ex-post to enhance the comparative advantage of the existing developmental mode of mass low-end production.

---

[1] The interview was conducted in Chinese. The author is responsible for all the transition from Chinese.

Before the discussion on the innovation pattern based on empirical results from the electronics industry survey, the overall innovative performance in Shenzhen and Dongguan would be displayed to gain a first insight into the development of regional innovation system.

5.5 Descriptive Profile of Innovation Activities in Shenzhen and Dongguan

Table 5.3 depicts the major statistics for each of the two cities. The population size in Shenzhen and Dongguan does not differ a lot, and the employment opportunities in both cities are also quite high. However, the industrial output value as well as the GDP (calculated as the value added) in Shenzhen is more than two times larger than that in Dongguan, which indicate a much higher productivity in Shenzhen than in Dongguan. Moreover, the pattern of specialization in high-tech sector in Shenzhen outstands from that in Dongguan in terms of industrial output value and employment.

As for the innovation indicators, Shenzhen's total R&D expenditure is more than six times higher than that in Dongguan, and the intensity of R&D investment is 3.4% for Shenzhen, which is comparable even to that in developed countries (OECD countries 2.3% in 2009, USA 2.9% in 2009, Japan 3.4% in 2009, Korea 3.3% in 2009; See OECD 2011). While in Dongguan, the intensity of R&D

Table 5.3 Major indicators in Shenzhen and Dongguan (2009). (Data sources: SMBS 2010; DMBS 2010; GPBS 2009)

	Shenzhen	Dongguan
Population	8,912,300	6,350,000
GDP (billion Yuan)	820	376
Industrial output value (billion Yuan)	1,582	676
% of High-tech manufacturing sector[a]	69%	39%
Employment	6,924,853	5,381,981
% of High-tech manufacturing and service sector[b]	33%	19%
Total R&D expenditures (billion Yuan)	27.97	4.14
% of GDP	3.4%	1.1%
R&D personnel	123,687	18,524
share of R&D personnel per 1000 employees	17.9	3.4

[a] High-tech manufacturing sector refers to ordinary equipment, special purpose equipment, transport equipment, electric equipment and machinery, telecommunications, computer and other electronic equipment (only above designated sized firms that include all state-owned firms and firm with over 5 million sales are calculated)
[b] High-tech manufacturing and service sector include the high-tech manufacturing sector above and service sector, i.e. information transfer, computer and software services, scientific research, technical services and geographical prospecting

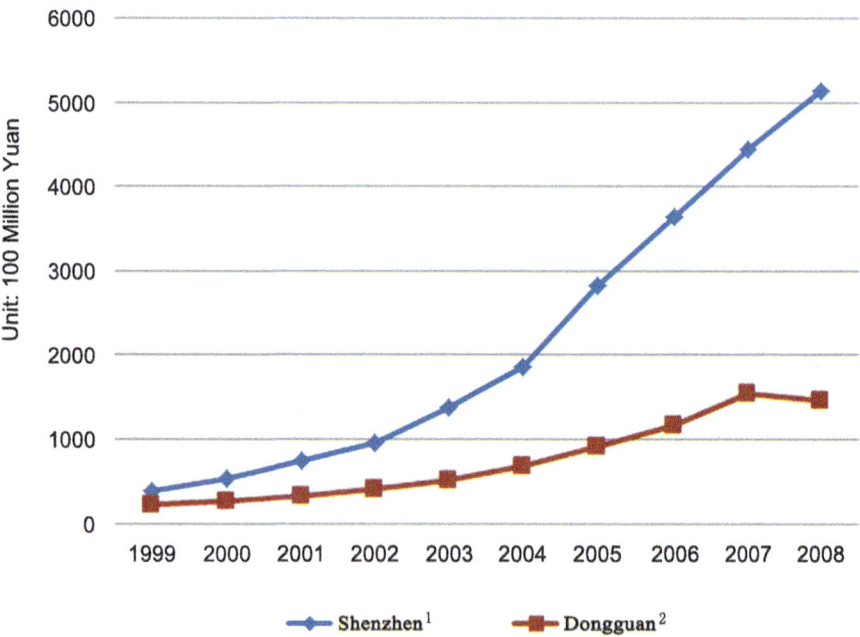

1. Output value in Shenzhen produced by high-tech firms with intellectual property
2. Output value in Dongguan produced by provincial-level high-tech firms

Fig. 5.3 Industrial output value of high-level high-tech companies (1999–2008). (Data source: SMBS (2000–2009, annual); DMBS (2009))

investment (1.1%) is still quite low, which is even lower than the national average level of 1.7% in 2009. Besides, the R&D personnel in Shenzhen outperforms that in Dongguan both in absolute and relative term, which all point to a higher level of human capital that enables the functioning of regional innovation system.

Because the high-tech manufacturing sector defined in Table 5.3 is very broad, firms in this sector might produce the same kind of products while possess different levels of technological capabilities. As a result, the performance of high-level high-tech companies in both cities is further investigated in Fig. 5.3. For Shenzhen firms, high-tech firms with intellectual property are defined as high-level high-tech firms. For Dongguan firms, it refers to provincial-level high-tech firms because firms earn this qualification only if they participate intensively in R&D activities or acquire scale economy in high-tech products. Thereby, these two measures are comparable in some way. Notwithstanding the Shenzhen definition is stricter than the Dongguan definition, the high-level high-tech companies in Shenzhen have been developing at a much higher rate than that in Dongguan during the period from 1999 to 2008. In 2008, the industrial output value produced by high-level high-tech companies in Shenzhen reached over 500 billion Yuan, which is more than three times higher than the output value in Dongguan. This comparing pattern demonstrates the stronger innovation capabilities for Shenzhen firms than that for Dongguan.

5.5 Descriptive Profile of Innovation Activities in Shenzhen and Dongguan

Table 5.4 Patent application and granted in Shenzhen and Dongguan. (Data sources: SMBS (1996, 2001, 2006, 2009); DMBS (2009))

		Shenzhen	Dongguan
1995	Patent application	1,104	325
	Patent granted	721 (65%)	262 (80%)
2000	Patent application	4,431	1,653
	Patent granted	2,401 (54%)	1,051 (63%)
2005	Patent application	20,940	6,694
	Patent granted	8,983 (43%)	1,974 (29%)
2008	Patent application	36,249	14,406
	Patent granted	18,805 (52%)	4,362 (30%)

Numbers in the parentheses are the passing rate of patent application

Table 5.5 Distribution of granted patent category in Shenzhen and Dongguan. (Data sources: SMBS 1996, 2001, 2006, 2009); DMBS (2009))

		Shenzhen	Dongguan
1995	Invent patent	7 (1%)	8 (3%)
	Utility model patent	280 (39%)	41 (16%)
	Design patent	434 (60%)	213 (81%)
2000	Invent patent	1 (–)	4 (–)
	Utility model patent	750 (31%)	344 (25%)
	Design patent	1650 (69%)	1051 (75%)
2005	Invent patent	917 (10%)	24 (1%)
	Utility model patent	3458 (38%)	1116 (36%)
	Design patent	4608 (51%)	1974 (63%)
2008	Invent patent	5409 (29%)	115 (1%)
	Utility model patent	7971 (42%)	3616 (45%)
	Design patent	5425 (29%)	4362 (54%)

Numbers in the parentheses are the share

The patenting activities further reveal the innovation capabilities of the business superstructure in both cities. Table 5.4 shows that the absolute number of patent application and patent grant in Shenzhen exceeds that in Dongguan to a large extent in the year 1995, 2000, 2005 and 2008. Although the passing rate of patent application is lower in Shenzhen than that in Dongguan in the year 1995 and 2000 (which might be also attributed to the low number of patent application in Dongguan), the passing rate overtakes that in Dongguan in spite of the much larger denominator (absolute number of patent application) for Shenzhen.

Among the granted patent in both cities, it is shown by Table 5.5 that the granted patents concentrate more on higher level of category such as utility model patent and invent patent in Shenzhen firms than that in Dongguan, Moreover, Shenzhen's concentration towards higher category compared to Dongguan is gradually strengthened from 1995 to 2008.

Table 5.6 Patent application of medium and large enterprises (2007–2008). (Data sources: SMBS (2008, 2009); DMBS (2009))

		Shenzhen	Dongguan
2007	Patent application	20,668	1,243
	Among which: inventions	15,322 (74%)	337 (27%)
2008	Patent application	22,391	1486
	Among which: inventions	15,053 (67%)	300 (20%)

Numbers in the parentheses are the share

For the patenting activities in private sector, Table 5.6 shows that the number of patent application of medium and large enterprises in Shenzhen is well above that in Dongguan to a great extent. In addition, the share of invention patents among the whole patent application is much higher for Shenzhen than that for Dongguan.

Overall, the comparison of the general innovation capabilities between Shenzhen and Dongguan suggests a much developed pattern of innovation capabilities in Shenzhen that would permit a well-functioning regional innovation system in which the interactive reciprocal innovation synergies is able to take place.

5.6 Empirical Evidence for Interactive Innovation

After comparing the divergent evolutionary paths of governance since the opening up and current innovation capabilities in Shenzhen and Dongguan, an empirical investigation into the scope and extent of interactive learning and systemic innovation in their primary industries, the electronics industry, was conducted in order to gain insights into the development of respective regional innovation systems. In the analysis, tobit regression was applied to examine the impact of external interaction with other business partners on firms' innovation outcomes.

Factor analysis was firstly applied to reduce the dimensions of independent variables in order to simplify the following regression. The derived factors are able to explain over 60% of the variance of the original sample. In order to avoid multicollinearity, seven variables were finally selected as the proxy variables for innovation behavior in the regression model, which is listed in Table 5.7.

Table 5.8 further demonstrates the control variables, covering firm characteristics and firm absorptive capacity.

The dependent variable in the regression is the average score of evaluation on the degree of improvement (ranging from 0 to 5 with increasing significance of change) on function expansion and categories upgrading. In Figs. 5.4, 5.5 and 5.6, the censoring pattern of the dependent variables for the whole sample, the Shenzhen sample and the Dongguan sample are shown respectively. Long (1997) demonstrates that OLS regression provides inconsistent estimates of the parameters when

5.6 Empirical Evidence for Interactive Innovation

Table 5.7 Selected variables for innovation behaviors

Indicators	Description
NPI_external partners	Interacting with *domestic customers, universities, research institutes and sales agents* to gain innovation ideas
NPI_internal efforts	Making *internal learning efforts* such as own ideas, license purchasing and reverse engineering
NPI_parent comp. & foreign	Relying on parent companies or foreign customers to gain innovation ideas
NPTK_active learning	Sending staff *to business partners* for training
NPTK_passive from customer	Receiving training and know-how from people sent *by domestic and foreign customers*
NPTK_passive from parent comp.	Receiving training and know-how from people sent *by parent company*
NPInteraction_informal	Interacting with innovation partners *within Guanxi networks*

Table 5.8 Control variables in the product innovation regression

	Indicators	Description
Firm characteristics	Size	Defined according to Chinese firm size standard, 1 as large firms with no less 300 million Yuan sales and no less than 2,000 employee, otherwise as small and medium-sized with the value of 0
	Ownership	1 as firms with foreign participation (wholly owned or joint venture), 0 as firms with 100% domestic participation
	Age	Years since establishment of the firm
Absorptive capacity	Educational level of technical staff	Proportion of technical staff with bachelor degree and above
	Initial product technology	Defined according to International Standard Industrial Classification of all Economic Activities, Rev 3[a], 1 as producing low-tech products when starting business, 2 as producing medium-tech products when starting business; 3 as producing high-tech products when starting business

[a] Specific classification of products into the different levels could be referred to Sect. 7.3 Appendix C

the dependent variable displays a censoring pattern. In this case, tobit regression was run in order to derive reliable estimation.

Table 5.9 tests the variation level between Shenzhen and Dongguan in terms of firm characteristic and absorptive capacity. In the surveyed sample, most of the firms are small and medium sized (94 and 89% in Shenzhen and Dongguan, respectively). The share of domestic firms in Dongguan is less than that in Shenzhen to a

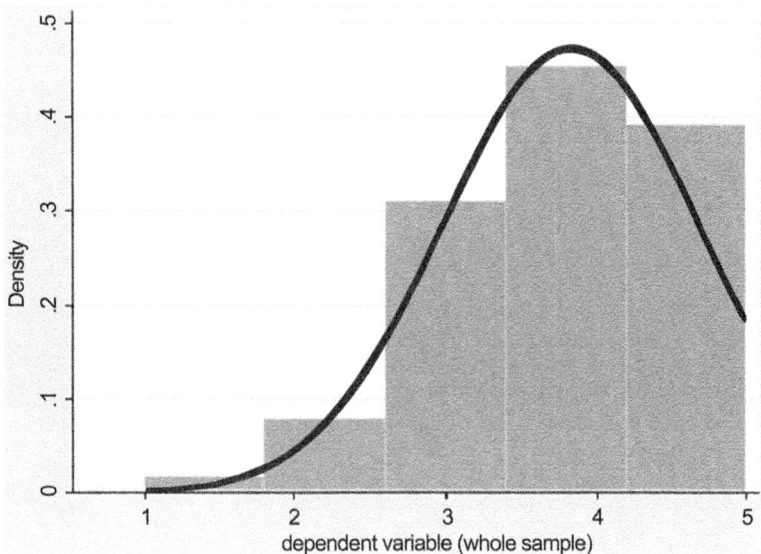

Fig. 5.4 Histogram distribution of product innovation outcome (Whole sample)

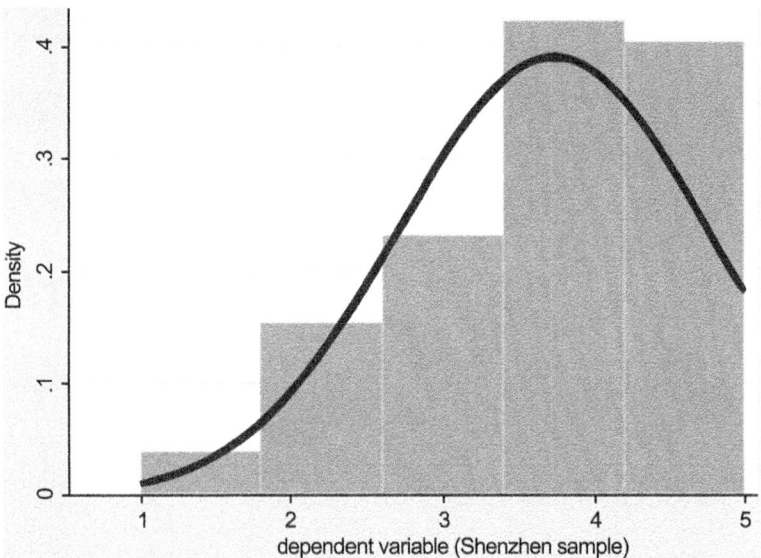

Fig. 5.5 Histogram distribution of product innovation outcome (Shenzhen sample)

significant degree. Technical staff possesses significantly higher absorptive capacity in Shenzhen than that in Dongguan according to the share of above bachelor degree technicians, and Shenzhen firms also start business in higher product technology than Dongguan.

5.6 Empirical Evidence for Interactive Innovation

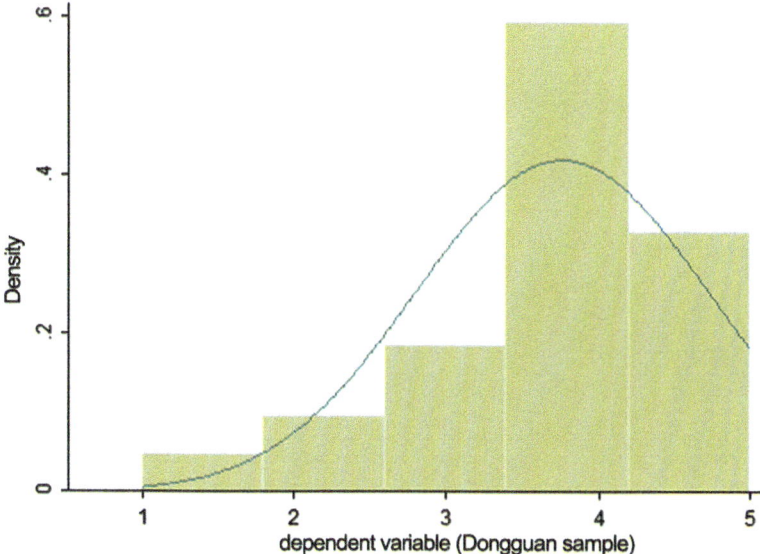

Fig. 5.6 Histogram distribution of product innovation outcome (Dongguan sample)

Table 5.9 ANOVA analysis in firm characteristics and absorptive capacity

Firm size (% of large firms)		*Shenzhen*	*Dongguan*
	Mean	0.06	0.11
	Sig.[1]	0.13	
Firm ownership (% of foreign firms)		*Shenzhen*	*Dongguan*
	Mean	0.28	0.47
	Sig.	0.001	
Firm age (years)		*Shenzhen*	*Dongguan*
	Mean	10.4	12.2
	Sig.	0.039	
Educational level of technical staff (%)		*Shenzhen*	*Dongguan*
	Mean	0.43	0.33
	Sig.	0.017	
Initial product technology		*Shenzhen*	*Dongguan*
	Mean	1.99	1.78
	Sig.	0.005	

In terms of innovation behavior, as shown in Table 5.10, Shenzhen firms turn more to external partners in triggering innovative ideas than Dongguan firms, but not at a significant level. On the other hand, Dongguan firms rely more on the transfer of tacit knowledge from parent companies and foreign customers, and more frequently use informal relations with friends and business partners.

Table 5.10 ANOVA analysis in innovation behavior

NPI_external partners		Shenzhen	Dongguan
	Mean	0.10	−0.07
	Sig.[1]	0.135	
NPI_internal efforts		Shenzhen	Dongguan
	Mean	0.02	0.11
	Sig.	0.427	
NPI_parent comp. & foreign		Shenzhen	Dongguan
	Mean	−0.22	0.27
	Sig.	0.000	
NPTK_active learning		Shenzhen	Dongguan
	Mean	−0.03	0.06
	Sig.	0.409	
NPTK_passive from customer		Shenzhen	Dongguan
	Mean	−0.02	0.10
	Sig.	0.31	
NPTK_passive from parent comp		Shenzhen	Dongguan
	Mean	−0.04	0.10
	Sig.	0.238	
NPInteraction informal		Shenzhen	Dongguan
	Mean	−0.14	0.14
	Sig.	0.014	

Table 5.11 shows the result of the tobit regression on product innovation outcome. Three models are run as a comparison: whole model pooling of the Shenzhen and Dongguan data, the Shenzhen model and the Dongguan model. All the models fit significantly better than an empty model, which is indicated by the significant level of the chi-square likelihood ratio. The whole model serves as an intermediary between the Shenzhen model and the Dongguan model, which reflects the difference between Shenzhen and Dongguan in a clearer way.

Observing firstly the variables indicating the behavior in the various stages of the product innovation process, Shenzhen firms combine their internal absorptive capacity with external interaction with other partners to trigger innovation ideas, which eventually boosts the innovation outcomes. In a regional innovation system, the interactive learning not only contributes to effective knowledge transfer, but also triggers the innovation, enabling capitalization on new creative resources from the complementary knowledge of various players in the cluster (Capello 1999). This indicates the strategy and capacity of Shenzhen electronics firms to capitalize on wider sources of knowledge spillover, including domestic customers, sales agents, universities and research institutes, which signify the maturing of the interactive regional innovation system in Shenzhen.

5.6 Empirical Evidence for Interactive Innovation

Table 5.11 Tobit regression on product innovation outcome

Independent variables		Product innovation outcome		
		Whole model	Shenzhen model	Dongguan model
NPI_external partners		0.31*** (0.091)	0.53*** (0.158)	0.12 (0.105)
NPI_internal efforts		0.20** (0.081)	0.39*** (0.135)	0.05 (0.093)
NPI_parent comp. & foreign		0.25*** (0.089)	0.21 (0.155)	0.25** (0.102)
NPTK_active learning		−0.05 (0.094)	−0.28* (0.147)	0.08 (0.118)
NPTK_passive from customer		−0.07 (0.087)	−0.43*** (0.135)	0.16 (0.103)
NPTK_passive from parent comp.		−0.08 (0.082)	−0.11 (0.133)	−0.12 (0.098)
NPInteraction_informal		−0.04 (0.083)	0.04 (0.140)	−0.07 (0.098)
Firm size		0.23 (0.276)	0.32 (0.522)	0.15 (0.305)
Ownership		−0.30* (0.153)	−0.53* (0.268)	−0.05 (0.206)
Firm age		0.008 (0.010)	0.03* (0.015)	−0.008 (0.013)
Educational level of technical staff		0.004* (0.002)	0.005 (0.003)	0.002 (0.003)
Initial Product technology	Medium tech vs. low tech[a]	0.15 (0.168)	0.08 (0.282)	0.16 (0.191)
	High tech vs. low tech[a]	0.37 (0.237)	0.14 (0.357)	0.60** (0.302)
Prob > chi2		0.0005	0.0006	0.0291
Pseudo R square		0.05	0.11	0.07
Number of observations		240	109	130

[a] Initial product as low tech as the default group, which means low tech as 0, the others as 1
Standard errors in parentheses; $*p<0.10$, $**p<0.05$, $***p<0.01$.

On the other hand, innovation ideas originating within strict hierarchical organizations, i.e. instructions from parent companies and foreign customers, boosts innovation outcome for Dongguan firms. Interactive learning in Dongguan is exclusively oriented to a fairly passive pattern of receiving orders to expand product functions and upgrade product categories from the organizationally proximate partners. Compared to the innovation activities in Shenzhen firms, the limited capacity for drawing upon a wider scope of external sources to foster innovation reflects the bottleneck of upgrading in Dongguan, where the internal absorptive capacity and external business environment do not permit the strategic use of interactive learning in the innovation process.

It is worth mentioning that the effect of either actively sending employees to learn technical experience or passively having engineers sent by other partners to pass on technical experience is significantly negative for Shenzhen firms. This might be related to the loss of technical staff in the process of gaining tacit knowledge. The higher absorptive capacity of the technical staff in Shenzhen firms than those in Dongguan enables them to absorb the knowledge from other organizations more effectively and also to identify the opportunities with higher salaries and positions. However, it should be cautiously interpreted because the labor mobility among local firms should contribute to effective interactive learning processes over the whole economy (Arrow 1962; Almeida and Kogut 1999). It is possible that firms gain the spilling-in human capital while losing others in the process of tacit knowledge transfer.

Comparison of the significance level of control variables for Shenzhen and Dongguan model further reveals some interesting points. For Shenzhen firms in the sample, older firms tend to have higher performance in product innovation. This variable demonstrates the long history of capability accumulation related to innovation activities, such as in technological development, management optimization and market research, contributes to higher absorptive capacity and higher effectiveness in bringing out better innovation results. In contrast, the small and insignificant impact of firm age on innovation outcome for Dongguan firms indicates that the firm strategy for accumulating technological and managerial capabilities around innovation activities is not conscious and systematic. However, Dongguan firms producing high-tech electronics products at the beginning, which indicates higher absorptive capacity, perform better than firms producing low tech electronics products at the beginning in a significant level of 90%. In short, firms in Dongguan rely more than Shenzhen firms on the routine accumulated gradually within the firm boundary, rather than on complementary knowledge outside the firm, leading to the lack of dynamism and incentive to trigger innovation. The innovation activities in Dongguan are rather passively led by global agents such as parent companies and foreign firms.

Post-estimation was run to ensure the robustness of the results. In Figs. 5.7, 5.8 and 5.9, the distribution of the residuals in the whole model, Shenzhen model and

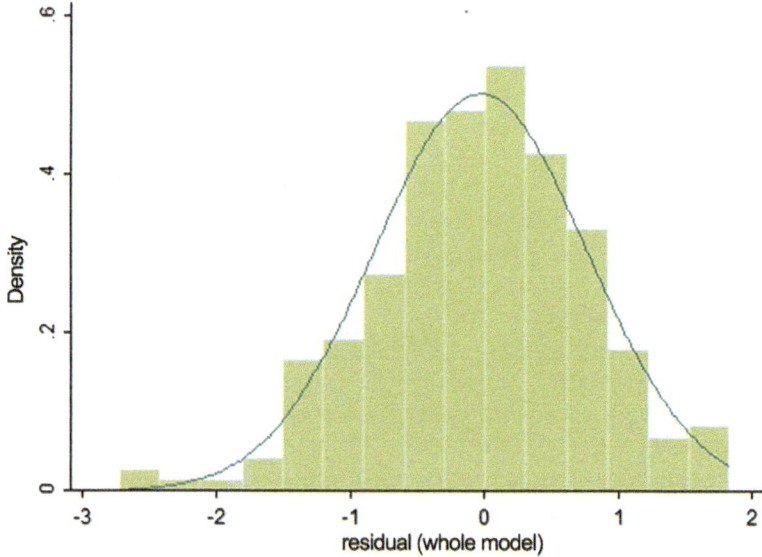

Fig. 5.7 Histogram distribution of model residuals (Whole Model)

5.6 Empirical Evidence for Interactive Innovation

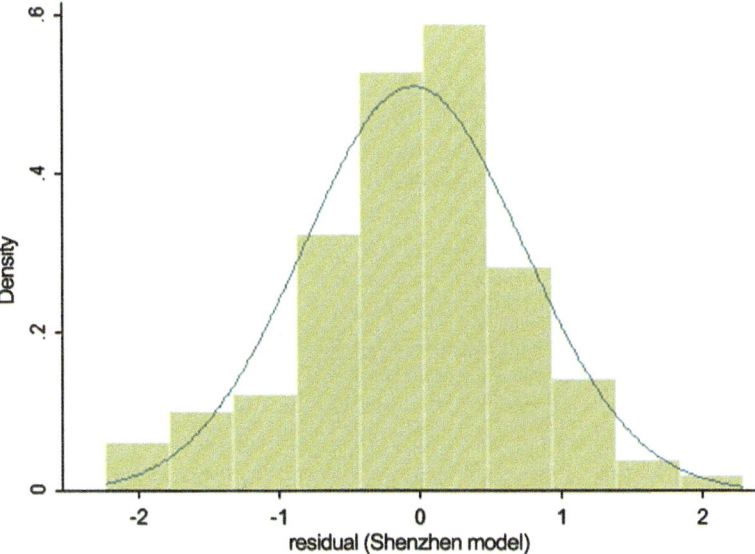

Fig. 5.8 Histogram distribution of model residuals (Shenzhen Model)

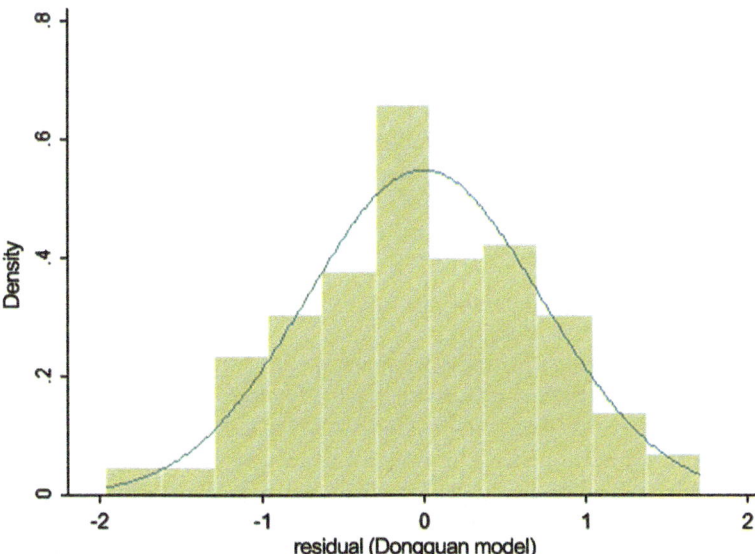

Fig. 5.9 Histogram distribution of model residuals (Dongguan Model)

Dongguan model were displayed in sequence. They all obey the normal rule, which indicates that heterokedastic issue, that might tortures the results of tobit model, does not exist.

5.7 Discussion and Conclusion

Governance perspective towards regional development and innovation is characterized as supply-side support, which aims to provide supportive resources, secure collective actions and establish the strategic goals (Hausner 1995). By comparing two cities from an evolutionary perspective, this chapter finds out that dirigiste governance modality in Shenzhen in the initial industrialization phase leads to a more mature and developed regional innovation system than the grassroots governance modality in Dongguan, though they both started the industrialization process in the wake of the opening policy in the late 1970s.

Insights from the empirical results show that dirigiste governance in the initial industrialization phase is more competent in providing innovation-related resources and adjusting the developmental path with strategic intervention than the grassroots governance, widening the scope of interactive learning and shaping the behavioral rationalities of firms to resort more to external complementary knowledge. While the newly recognized strand of grassroots governance supports its competency to mobilize the local resources and interdependencies (Amin 2002), the result suggests a rather contrasting pattern, indicating that this approach in the initial phase of industrialization might lead to a negative lock-in effect in the face of restructuring and upgrading by restricting the firms within the repeated and narrow path of knowledge accumulation and generation.

As evolutionary investigation is subject to context, it should be remembered that the two cities in this study started the rapid industrialization out of nothing—that is to say—with a barren endowment of local skills and industrial base. In this case, the grassroots approach tends to restrain the scope of development within the disposal of less competent local authorities. On the other hand, the empirical findings on the success of dirigiste governance in shaping innovative synergies should not be viewed as arguments favoring the central planning method of development in Keynesian legacy. In fact, this institutional advantage is combined within the market opportunity brought by the reorganization of global production networks, boosting a plurality of autonomous decision-making agents with respective strategic goals in the transition economy.

Grassroots governance in China has been widely applied since it was cost efficient for the central government and has actually mobilized the initiative of local governments to develop the economy. For clusters that developed out of grassroots governance in the early phase of industrialization, two lessons can be learned to boost the development of the regional innovation system. Firstly, strategic planning of industrial development should be carried out to avoid negative lock-in, adjusting the developmental path to meet the changing market environment in time and identifying related new industries. Most importantly, levels of governance should be accordingly regulated and balanced to unfasten the vested interests aiming for deviating development goals. Secondly, policy focus should be put upon enhancing the absorptive capacity of firms and related organizations, such as attracting high quality human capital and encouraging the conscious accumulation and development of technological capabilities within firms.

As for the more developed case in Shenzhen, experiences of building a sustainable regional innovation system can be further learned from Europe and the USA where innovation governance models are mature. In Europe, economic agents depend to a significant degree on public institutions for fostering innovation activities, while in the USA, the role of private institutions, e.g. banks and venture capitals, is prominent in organizing systematic learning and innovation. Ultimately, if the supporting governance is competent in performing inclusive, monitoring, consultative and networking features, it is more likely to allow high potential regional innovation systems.

The comparative study between Shenzhen and Dongguan, China captures the governance modality in the initial industrialization phase and its evolution with market change as an important factor that leads to different degrees of regional innovation system development. As indicated by Cooke (2004, pp. 17), the development of regional innovation systems are context-specific, and its dynamics goes along with globalization, the rise of knowledge industries and the hollowing-out of traditional industries. Therefore, it would be useful to identify the elements of governance in relation to the business needs under new market trends. Furthermore, more thoughts should be put into the question of how to keep the dynamics and prevent the inertia of governance modality in the face of change and transformation.

References

Almeida P, Kogut B (1999) Localization of knowledge and the mobility of engineers in regional networks. Manag Sci 45(7):905–917

Amin A (1999) An institutionalist perspective on regional economic development. Int J Urban Reg Res 23(2):365–378

Amin A (2002) Spatialities of globalisation. Environ Plan A 34(3):385–400

Arrow KJ (1962) The economic implications of learning by doing. Rev Econ Stud 29(80):155–173

Asheim BT, Coenen L (2005) Knowledge bases and regional innovation systems: comparing nordic clusters. Res Policy 34(8):1173–1190

Asheim BT, Isaksen A (2002) Regional innovation systems: the integration of local 'sticky' and global 'ubiquitous' knowledge. J Technol Transf 27(1):77–86

Boschma R (2004) Competitiveness of regions from an evolutionary perspective. Reg Stud 38(9):1001–1014

Boschma R, Iammarino S (2009) Related variety, trade linkages, and regional growth in Italy. Econ Geogr 85(3):289–311

Braczyk HJ, Cooke P, Heidenreich M (1998) Regional innovation systems: the role of governances in a globalized world. Routledge, London

Capello R (1999) Spatial transfer of knowledge in high technology milieux: learning versus collective learning processes. Reg Stud 33(4):353–365

Cohen WM, Levinthal DA (1990) Absorptive capacity: a new perspective on learning and innovation. Adm Sci Q 35(1):128–152

Cooke P (1992) Regional innovation systems: competitive regulation in the new Europe. Geoforum 23(3):365–382

Cooke P (2001) Regional innovation systems, clusters, and the knowledge economy. Ind Corp Change 10(4):945–974

Cooke P (2004) Regional innovation systems: an evolutionary approach. In: Cooke P, Heidenreich M, Braczyk HJ (eds) Regional innovation systems: the role of governance in a globalized world, 2nd edn. Routledge, London, pp 1–18

Cooke P, Gomez Uranga M, Etxebarria G (1997) Regional innovation systems: institutional and organisational dimensions. Res Policy 26(4–5):475–491

Cooke P, Gomez Uranga M, Etxebarria G (1998) Regional systems of innovation: an evolutionary perspective. Environ Plan A 30:1563–1584

Cooke P, Heidenreich M, Braczyk HJ (eds) (2004) Regional innovation systems: the role of governances in a globalized world. Routledge, London

Dalum B, Johnson B, Lundvall B (1992) Public policy in the learning economy. Printer, London

DMBS (Dongguan Municipal Bureau of Statistics) (1995–2010) Dongguan Tongji Nianjian (Dongguan statistical yearbook). China Statistics Press, Beijing

DMBS (Dongguan Municipal Bureau of Statistics) (2009) Dongguan Shehui Keji (Dongguan society and technology). Dongguan, China

Doloreux D (2002) What we should know about regional systems of innovation. Technol Soc 24(3):243–263

Doloreux D (2004) Regional innovation systems in Canada: a comparative study. Reg Stud 38(5):479–492

Easterly W (2008) Institutions: top down or bottom up? Amer Econ Rev 98(2):95–99

Eshag E (1991) Successful manipulation of market forces: case of South Korea, 1961–78. Econ Polit Wkly 26(11/12):629–644

Gambardella A (1992) Competitive advantages from in-house scientific research: The US pharmaceutical industry in the 1980s. Res Pol 21(5):391–407

GECC (Gungdong Electronics Chamber of Commerce) (2002) Guangdong Electronics Yearbook. http://www.guangdongdz.com/4c_of_c/annual.asp. Accessed 31 Nov 2012

Goldsmith AA (2007) Is governance reform a catalyst for development? Governance 20(2):165–186

GPBS (Guangdong Provincial Bureau of Statistics) (2009) Guangdong Dierci Quanguo R & D Ziyuan Qingcha Tongji Ziliao Huibian (Guangdong province of the second national statistics compiled by the R & D resources inventory). Guangdong Statistics Press, Guangzhou

Haggard S (2004) Institutions and growth in East Asia. Stud Comp Int Dev 38(4):53–81

Hausner J (1995) Imperative vs. interactive strategy of systemic change in central and Eastern Europe. Rev Int Polit Econ 2(2):249–266

Howells JRL (1999) Regional Systems of Innovation? Cambridge University Press, Cambridge

Hudson R (1994) Institutional change, cultural transformation, and economic regeneration: myths and realities from Europe's old industrial areas. In: Amin A, Thrift N (eds) Globalization, institutions, and regional development in Europe. Oxford University, Oxford, pp 196–216

Jacobs J (1969) The economy of cities. Random House, New York

Kim L (1999) Building technological capability for industrialization: analytical frameworks and Korea's experience. Ind Corp Change 8(1):111–136

Könnölä T, Unruh GC, Carrillo-Hermosilla J (2006) Prospective voluntary agreements for escaping techno-institutional lock-in. Ecolog Econ 57(2):239–252

Leonard-Barton D (1992) Core capabilities and core rigidities: a paradox in managing new product development. Strateg Manag J 13(S1):111–125

Leung CK (1993) Personal contacts, subcontracting linkages, and development in the Hong Kong-Zhujiang Delta Region. Ann Assoc Am Geogr 83(2):272–302

Levitt B, March JG (1988) Organizational learning. Ann Rev Sociol 14:319–340

Long JS (1997) Regression models for categorical and limited dependent variables. Sage, Thousand Oaks

Lundvall BA (1992) National innovation systems: towards a theory of innovation and interactive learning. Pinter, London

Luthje B (2004) Global Production networks and industrial upgrading in China: the case in electronics contract manufacturing. Paper presented at the international conference on Multinationals in China-Competition and Cooperation. 9–11 July 2004, Sun Yat-sen University, Guangzhou, China

Martin R (1999) Critical survey. The new 'geographical turn' in economics: some critical reflections. Camb J Econ 23(1):65–91

Martin BR, Johnston R (1999) Technology foresight for wiring up the national innovation system: experiences in Britain, Australia, and New Zealand. Technol Forecast Soc Change 60(1):37–54

Morgan K (2004) The exaggerated death of geography: learning, proximity and territorial innovation systems. J Econ Geogr 4(1):3–21

Organization for Economic Co-operation and Development (OECD) (2011) Main Science and Technology Indicators database. OECD, Paris

Peng MW, Heath PS (1996) The growth of the firm in planned economies in transtition: institutions, organizations, and strategic choice. Acad Manag Rev 21(2):492–528

Porter ME (1998) Clusters and the new economics of competition. Harv Bus Rev 6:77–92

Qian Y, Roland G (1998) Federalism and the soft budget constraint. Am Econ Rev 88(5):1143–1162

Revilla Diez J (2000) The importance of public research institutes in innovative networks-Empirical results from the metropolitan innovation systems Barcelona, Stockholm and Vienna. Eur Plan Stud 8(4):451–463

SECC (Shenzhen Elecronics Chamber of Comerce) (2004) Shenzhen Electronics Yearbook. http://www.guangdongdz.com/special_column/sznj/zl02.asp. Accessed 30 Nov 2012

SMBS (Shenzhen Municipal Bureau of Statistics) (1995–2010) Shenzhen Tongji Nianjian (Shenzhen Statistical Yearbook). China Statistics Press, Beijing

Staber U (2001) The structure of networks in industrial districts. Int J Urban Reg Res 25(3):537–552

Stam E, Lambooy J (2012) Entrepreneurship, knowledge, space, and place: evolutionary economic geography meets Austrian economics. In: Anderssond E (ed) The spatial market process (Advances in Austrian Economics Vol 16). Emerald, Bingley, p 81–103

Storper M, Harrison B (1991) Flexibility, hierarchy and regional development: the changing structure of industrial production systems and their forms of governance in the 1990s. Res Policy 20(5):407–422

Sweeney GP (1991) Technical culture and the local dimension of entrepreneurial vitality. Entrep Reg Dev 3(4):363–378

Tripsas M (1997) Unraveling the process of creative destruction: complementary assets and incumbent survival in the typesetter industry. Strateg Manag J 18(summer special issue):119–142

Yang C (2012) Restructuring the export-oriented industrialization in the Pearl River Delta, China: Institutional evolution and emerging tension. Appl Geogr 32(1):143–157

Yeung HWC (2000) Local politics and foreign ventures in China's transitional economy: the political economy of Singaporean investments in China. Polit Geogr 19(7):809–840

Zahra SA, George G (2002) Absorptive capacity: a review, reconceptualization, and extension. Acad Manag Rev 27(2):185–203

Chapter 6
Conclusions

Abstract The final chapter summarizes the key findings in the study and draw some constructive policy implications therefrom. Echoing with the research questions posed in Chap. 1, the study has firstly provided potent evidence of the triggering effect of FDI on the regional innovation performance, which demonstrated the role of external knowledge spillover in the formation of a regional innovation system in latecomer regions. Apart from meso-level evidence, the study has utilized the first-handed survey data to demonstrate firm-level evidence for the positive role of interactive learning, supported by informal *Guanxi* with business partners, relatives and friends, in promoting the innovation outcomes for the electronics firms. However, unstable role of social proximity in innovation is shown, suggesting the absence of supported institutions and related organizations for complex and highly risky innovation activities. By an inter-city comparative case study, the study has uncovered the path-dependent nature of initial governance modality for shaping a mature regional innovation system. Future research direction concerns the theoretical and empirical endeavors for other mechanisms of distributive system on the regional level and the negative effect of informal *Guanxi* network in innovation, as well as the methodological issues in collecting survey data. Having recognized the systematically weak performance of market-based systems, the study ends with three domains of policy interventions that can be undertaken to strengthen the distributive power of the innovation system at the regional level.

6.1 What do We Learn About the Chinese Regional Innovation System?

The term 'regional innovation system' has a different meaning in China, since innovation here is characterized mainly as an exploiting process of externally sourced knowledge, rather than knowledge generation. In the preceding chapters, the pattern of interactive learning and systemic innovation, which is closely related to the formation of a regional innovation system, has been investigated among the electronics firms in the PRD, China. Furthermore, the spatial differences in the region have been uncovered with an evolutionary observation of the governance infrastructure that incentivizes and supports the innovation activities in the business sector. To summarize the insights from the empirical investigation in the PRD, China, answers will be provided to the key research questions formulated in Chap. 1 in the following three blocks.

6.1.1 The Shaping of Interactive Learning Behavior and Systemic Innovation

In the third wave of globalization in 1980s, the bulk of foreign capital flowed to developing countries in the form of direct investment in manufacturing (Dollar 2005). During this process, the PRD region was integrated into the global economy following the introduction of the opening policy in 1978. However, the production systems in these regions are strongly governed by the global lead firms and global buyers (Humphrey and Schmitz 2008; Yeung 2009). The FDI-dependent mode has provided great impetus for the rapid industrialization in these regions.

The central question for the innovation and upgrading in China is: how and under what circumstances do knowledge spillovers sourced externally trigger knowledge spillovers on the local scale, enabling the formation of regional innovation systems in latecomer export-oriented regions (T1)? This theoretical question relates to the first empirical question E1 "Have local-scale knowledge spillovers come into being to sustain long-term development in the face of a changing and fragile post-crisis global market in the export-oriented Guangdong Province, China?". As demonstrated by the analysis and empirical evidence in Chap. 2, external knowledge spillover has triggered the knowledge spillover within industries mainly through the channel of FDI in Guangdong Province. FDI facilitates the function of local knowledge spillover channels by enabling inter-firm collaboration and cooperation, enhancing human capital and accelerating spin-off activities. The result further shows that the impact of FDI stock on new product development relies on a diversified urban economy rather than a specialized one. On the other way around, knowledge spillover within the same industry, despite its role in promoting the learning by doing and knowledge exploitation processes within the domestic sector, does not interact with FDI and import effect in the region. This suggests the limit of highly specialized spatial economy in the PRD, in particular when the absorptive capacity is low and specific know-how in the region is little to attract foreign firms to interact with.

The knowledge spillover effects produced through interactive learning either with global partners or domestic partners are further supported by the firm-level investigation in Chap. 3 and 4. In Chap. 3, the motive of interactive learning is firstly theoretically discussed, which is elaborated upon by the T2 question "Why do firms undertake interactive learning with external partners in the decision-making and implementing process of innovation activities?" Following the argument of organization routine proposed by Nelson and Winter (1982), the bounded rationality and competence of firms necessitates the complementary use of interactive learning with other organizations in the aspect of searching for relevant information in order to make innovation-related decisions, as well as of seeking the support of codified and tacit knowledge in the innovation process. From the empirical results in Chap. 3, it is concluded that a wider scope and higher intensity of interactive learning promotes the innovation outcomes for the electronics firms in the PRD, which provides the answer to the first part of E4 "What is the effect of interactive learning in general on innovation outcomes?" In addition, the empirical investigation in Chap. 4 further discovers that socially active innovators outperform lame innovators.

In Chap. 4, an attempt is made to relate firms' internal absorptive capacity to their innovating behavior. The empirical question E2—"Which aspects of absorptive capacity enable the electronics firms to undertake interactive learning with external partners through strategies of using organizational proximity and social proximity in product innovation processes?"—is then addressed. It is shown that the production experiences in high-tech fields prepare the firms with more capability and resources to undertake interactive learning using either social proximity or organizational proximity. Nevertheless, parameters for R&D activities, including the presence of technology centers as well as the possession of design capability and development capability, do not differ significantly for interactive learning groups from the lame innovators. This result does not support Cohen and Levinthal's (1990) finding on R&D's function in creating and exploiting new knowledge. However, it should be pointed out that in their research, R&D refers to more basic research that is able to prepare the firms with general background knowledge to exploit new scientific knowledge. The concentration of test and development activities in the R&D function among the electronics firms in the PRD might be able to explain the incapability of R&D to boost the absorptive capacity of firms to undertake interactive learning for new knowledge exploitation. In any case, the results on the effect of R&D should be interpreted consciously, and this issue will be further addressed in the discussion on the limitations of the work.

6.1.2 The Informal Aspect of Innovation Activities in China

Under the proximity concept of investigating interactive learning in the PRD, China, the informal aspect of innovation activities has been covered in the analysis in Chap. 3 and 4. In the context of China, informal interaction among various economic players and the embeddedness in "*Guanxi*" networks are important factors for doing business. In the first phase of the DFG research project, on which this work is developed, the informal aspect of economic life in China has been investigated with an analytical focus on production activities. Meyer (2011) has conducted a thorough investigation into the informal interaction mode of customer-producer relations with regard to achieving a high level of flexible production, concluding that informality contributes to finding new customers or producers quickly, increasing speed to market and saving time when conflict emerges in contractual enforcement. This book further develops the understanding of the informal aspect of economic life in China with respect to innovation activities, which is of great relevance to the upgrading issue in the face of rising factor prices, intensifying competition from other regions and shrinking foreign markets.

Empirical results in Chap. 3 verify that informal relationships with business partners, relatives and friends are a widely applied practice in interactive learning during the product innovation process. The firms that undertake the widest scope and highest intensity of interactive learning activities tend to apply informal *Guanxi* networks to assist in searching and contacting with the related business partners, from whom reliable information and knowledge can be sourced from in the product innovation process.

In Chap. 4, the insights into the informal aspects of interactive learning are further elaborated upon explicitly under the concept of social proximity. It firstly probes into the question T3: "What is the role of social proximity and organizational proximity in interactive learning activities in latecomer export-oriented regions?" Based on the review of the global production network literature, it hypothesizes that organizational proximity with foreign parent companies and foreign customers is helpful in transferring tacit knowledge for firms in latecomer countries in the early phase of development, and enables firms to foster innovation further with a more sophisticated supply chain and sufficient absorptive capacity. On the other hand, social proximity, embodied by the informal *Guanxi* asset in China, is able to play a role in trust building by shaping local dynamic innovative synergy with the precondition of mature internal absorptive capacity. Through the comparative analysis of the extent to which social proximity with local partners and organizational proximity with global partners foster product innovation, it is shown that electronics firms in the PRD, China, resort more to social proximity than to organizational proximity in interactive learning processes (answer to E3: "How is interactive learning organized in the burgeoning regional innovation system? To be more specific, does interactive learning embed more in socially proximate networks or organizationally proximate networks?"). Nevertheless, socially active firms do not outperform organizationally dependent firms in terms of transforming the proximity assets into better innovation outcomes (answer to the second part of E4: "And what is the effect of interactive learning embedded within socially proximate networks and organizationally proximate networks on innovation outcomes respectively?").

This result suggests an unstable role of social proximity, embodied by the *Guanxi* network, in sustaining trust with regard to innovation activities with high levels of uncertainty and risk. As pointed out by Chesbrough and Teece (1996), information sharing can be reduced or biased as each seeks to get the most at the other's expense. Therefore, the effect of informal methods, such as *Guanxi* networks in the Chinese business mode on promoting systemic innovation among firms needs to be assured by a durable and time-consuming relationship construction under the market circumstances, where the interests of the interacting and cooperating firms are kept in correspondence and harmony. Therefore, supported institutions and related organizations that shape common norms and rules should be in place to sustain stable interactive learning processes.

6.1.3 The Spatial Difference under Divergent Governance Infrastructure Evolution

Regional innovation systems stress the role of governance infrastructure in supporting the business superstructure of innovation activities. A trend towards the evolutionary perspective on regional innovation systems has been testified to by the republication of the seminal book "Regional innovation systems: the role of governance in a globalized world" in 2004 (Cooke et al. 2004). Compared to the first edition (Braczyk et al. 1998), this book analyzes the evolutionary path of regional innovation systems with changing contextualization elements such as globalization, the rise of the knowledge economy and the deindustrialization process. Chapter 5 adjusts the evolutionary lens according to the context of China, where the regional innovation system is being incubated from the production system relying heavily on integration into the lower end of global production networks but presently facing rising factor prices and upgrading pressure, and focuses the investigation on the transition from governance that supports initial industrialization to governance that supports the innovation activities.

By means of a thorough theoretical discussion of governance infrastructure, both in production systems and innovation systems as well as their relationships, Chap. 5 depicts the evolutionary path of the globalized grassroots production system and the globalized dirigiste production system towards the regional innovation system, providing an answer to question T4: "What leads to the dynamics and inertia of regional innovation system under different governance infrastructures?". It is concluded that globalized grassroots production systems with a weak industrial base tend to encounter competency traps and complex vested interests, leading to the risk of negative lock-in and sticky inertia along the evolutionary path towards the one supporting innovation systems. Compared with the globalized grassroots production system, the globalized dirigiste production system, which is mostly initiated and governed by national level agencies with more power, is more capable of injecting new dynamics into development and accumulating the skill stock from the central assignments, for example through relocation of large state-owned firms and research institutes. Moreover, the awakening of market mechanisms to facilitate the transition of dirigiste governance to network governance is also strengthened as one of the most important determinants in the dynamics of the dirigiste approach.

The inter-city comparative study between Shenzhen and Dongguan electronics firms yields the answer to question E5: "How do regional innovation systems in Shenzhen and Dongguan, China differ from each other in the scope and effect of interactive learning, considering that the two cities are evolving towards regional innovation systems under different governance infrastructures in the initial industrialization phase?". It demonstrates that electronics firms in Shenzhen are able to capitalize on a wider scope of interactive learning activities to foster innovation, while Dongguan electronics firms have restricted the scope of learning to within the hierarchical boundary with parent companies and foreign customers, leading to the reliance on the transfer of tacit knowledge from a limited number of players to foster innovation. Combined with the answer to question T4, it can be concluded that the dirigiste approach in globalized production systems without established industrial base is more competent in providing innovation-related resources and adjusting the developmental path with strategic intervention than grassroots governance, widening the scope of interactive learning and shaping the behavioral rationalities of firms to resort more to external complementary knowledge.

6.2 Directions of Future Research

This section reflects upon the limitations of the study and provides directions for future research accordingly. Limitations and the corresponding research directions can be summarized under the following three aspects.

6.2.1 The Mechanism of Distributive System on the Regional Level

This chapter analyzes the distributive power of China's regional innovation system with the focus on inter-organizational interactive learning. Nevertheless, distributive channel of information and knowledge in the regional innovation system are not only sustained through inter-firm vertical collaboration and horizontal cooperation, but also through labor mobility and spin-off activities.

It is commonly emphasized that regional innovation systems depend on the density and quality of the network among firms, knowledge-related organizations and institutions. However, the issue of how labor and its related governance contribute to the systems of innovation processes is still neglected in the literature of regional innovation system. In small firm district, the short and turbulent life of small firms results in a high turnover of qualified workers. According to the firm interviews in the pre-test phase of the survey in this study, many firms rely on talent poaching to ensure the success of innovation projects. Further research on the learning by hiring effect within the regional innovation system in China could go into two directions. Firstly, the factors that lead to firms' reliance on interaction with individuals instead of organizations as a way to gain complementary knowledge can be explored, as unbalanced absorptive capacity among the latecomer firms is likely to disturb the sustainability of inter-organizational interaction. Secondly, the role of labor-related governance, such as training agencies, employment agencies, labor unions and supported laws and practice in promoting or hindering the innovation activities should be explored further.

Another factor that contributes to the distributive power of innovation systems is the spin-off activities. Parhankangas and Arenius (2003) define three types of spin-offs as spin-offs developing new technologies, spin-offs serving new markets and restructuring spin-offs. Although these spin-offs differ from each other in the intensity of resource sharing linkages and knowledge transfer with the parent companies, the effect of interactive learning among them is assumed to be higher due to the sharing of experience and routines. Asheim and Coenen (2005) also point out that spin-off activities are key methods of knowledge application and exploitation, especially in scientific knowledge-based clusters. Therefore, further research should investigate the sustained relationship and interaction between spin-offs and the incubators and parent firms, as well as their role in exploiting and commercializing the new combination of knowledge.

6.2.2 The Negative Effect of Informal Guanxi Network on Innovation

In the theoretical discussion on the role of *Guanxi* networks in innovation in Chap. 4, the downside of *Guanxi* networks in fostering innovation has been touched upon. However, this book has not yet undertaken an empirical investigation into the negative effect of informal *Guanxi* networks in innovation due the limits of the firm survey resources.

The disadvantage of *Guanxi* lies in two aspects. Firstly, it can damage the development of firm internal capability due to limits of time and resources. It is an intricately woven interpersonal network that requires constant monitoring, investment and subtle utilization. Gains in terms of *Guanxi* network improvement must result in lack of investment in other aspects such as managerial capability and technological capability. Secondly, the rent-seeking kind of *Guanxi* network can harm the overall efficiency of economies. Resources are distributed according to *Guanxi* with government officials, rather than according to capability and efficiency of the firms. This actually suppresses the firms' incentives to invest in long-term technological capability accumulation, leading to underdevelopment of absorptive capacity that hampers the effect of interactive learning on innovation activities.

Therefore, absorptive capacity constitutes the primary leveraging tool in the capacity to capitalize on informal *Guanxi* networks with regard to innovation outcomes. Only when the firms develop enough capacity to absorb, adapt and exploit the information and new knowledge, can *Guanxi* networks contribute to the innovation outcome of the system at the regional level. Future research should further collect the empirical evidence of the negative effect of informal practices on innovation activities and compare them to their positive role in shaping trust and reducing risk in the interaction process.

6.2.3 Methodological Issues in the Survey Design

Most of the empirical investigations of this book are based on an electronics firm questionnaire survey conducted at the end 2009. Following the empirical investigation into the questions concerned with innovation activities, there are two primary aspects in which improvement can be made in future firm survey investigations in regions such as the PRD, China.

Firstly, a small sample size of serious R&D undertakers in the PRD hinders the systematic investigation of the effect of R&D intensity on boosting the absorptive capacity and shaping interactive learning behavior. Especially in the questionnaire survey, it was very difficult to obtain comparable and accurate data on R&D activities. Many firms tended to overrate the presence of the R&D function and the intensity of R&D expenditures, since R&D is still a trendy word in China and firms are not informed as to the exact meaning of it. Sometimes even minor adaptation activities are viewed as R&D, which leads to incomparability with firms that undertake true R&D functions. Therefore, improvements to future firm surveys should be made in the following respects: firstly, a detailed investigation should be conducted to differentiate between basic research, applied research and test & development activities among the general R&D functions. Most importantly, efforts should be devoted during the survey to explaining the true meaning of each element of R&D activities to the surveyed firms, attempting to gather a more comparable dataset out of it. In short, the book raises a call for a more thorough investigation of R&D activities in the Chinese context with regard to the composition and quality of R&D activities measurement.

Secondly, the interaction mode with each business partner in each specific innovation process has not been identified under this survey design. The matrix of questionnaires would otherwise become too complex for the firms to answer. In order to ensure the success of the survey, only general information on interaction mode with business partners is identified. As this book constitutes experimental exploration with regard to the role of informal *Guanxi* networks in innovation activities, future empirical research design should select the focus according to the respective research interests and investigate further in which process of innovation (e.g. collecting information and gaining support of equipment and technical experience) and with which partners (e.g. parent companies, foreign customers, domestic customers, suppliers, universities and other related organizations) firms apply informal networks to foster the innovation outcomes.

6.3 Policy Implications

The systemic innovation approach leads to different policy focus compared to the linear innovation approach (Smith 2000). The linear innovation approach points out the market failures in the knowledge generation process, in which the policy should intervene in order to reach an optimal production of knowledge stock. Taking this into consideration, policy intervention should focus on encouraging the indigenous knowledge production process, such as R&D activities, through the channel of subsidies and intellectual property rights protection (Arrow 1962). On the other hand, the systemic innovation approach does not necessarily exclude actions of this kind, and further underlines the systematically weak performance of market-based systems. In short, policy intervention following the logic of the system approach aims to encourage the sharing and joint exploitation of knowledge stocks in the process of new knowledge commercialization.

6.3 Policy Implications 163

This book adopts a systemic approach towards the innovation issue in China and highlights the role of interactive learning in the formation and development of the regional innovation system in the PRD, China. Based on the findings in the previous chapters, the following suggestions on policy intervention can be made for enhancing the innovation capability of firms and strengthening the distributive power of the innovation system at the regional level. These three aspects of policy recommendation provide illuminating answers to question P1: "What policy implication can be drawn from the previous answers from the theoretical and empirical perspectives in order to enhance the innovation capability of firms and regions in China?"

6.3.1 Enhancing and Balancing the Firm Absorptive Capacity

Absorptive Capacity is the prerequisite for undertaking interactive learning. Moreover, an equivalent stock of knowledge and capabilities between firms ensures the sustainable innovative synergies based on reciprocal exchange. Otherwise, firms tend to restrict the knowledge flow within the firm boundary in order to avoid opportunist behavior from other firms.

In order to shape the reciprocal regional innovative synergies, policy actions can intervene in the following three respects. Firstly, the government—especially municipal governments—should devote resources to city image management and create high-quality living environments, attracting highly qualified technical and managerial talents. According to the results in Chap. 4, the focus on talent attraction can be put on the educated managerial talents as well as entrepreneurs with overseas background to strengthen the strategic coupling with global partners in promoting upgrading and innovation. Measures such as setting up business incubators for returned talents can be strategically implemented. Moreover, related labor agencies should cooperate closely with the firms to identify the firms' demand for skill and organize corresponding recruitment activities. Secondly, government at the various levels should focus on the attraction of high-tech operations and encourage firms to upgrade to high-tech fields. As demonstrated by the results in Chap. 4, production experience and current practices in high-tech fields of the electronics industry is most effective in boosting the absorptive capacity and eventually promoting interactive learning among the firms. Last but not least, government should provide incentive for firms to keep sequential records of technological programs and exchanges, initiating the accumulation of technological capability at the firm level. This aim can be achieved through measures of specific funding programs.

6.3.2 Identifying and Supporting the Capacity of Interactive Learning

The results derived from the empirical investigation in Chap. 3 and 4 affirm the positive role of interactive learning in fostering innovation for electronics firms in the PRD, China. However, the empirical results also reveal that only a limited number of firms are capable of capitalizing on interactive learning in product innovation processes. The underdevelopment of interactive learning activities can be attributed not only to the underdevelopment of internal absorptive capacity to understand and adapt the external knowledge, but also to the limited resources for identifying related knowledge and searching for appropriate partners. In the study, it is shown that firms can use proximity to identify the interacting partners in product innovation processes. However, the dynamic proximity building process should be matched with certain relatedness and complementation of knowledge to share and inspired with both sides.

As interactive learning is a process of approaching externalities without firm boundaries, public policy can play a role in assisting the firms to establish proximity with related partners in the interactive learning activities. Smith (2000) suggests that the identification of the large externalities should be central to policy formation and operation. Policy actions in this respect include:

- Firstly, identify the direct and indirect knowledge inputs for a sector in order to enhance the understanding of both the depth and complexity of knowledge bases that are relevant in the regional innovation system (Smith 2000).
- Provide sources of generic scientific technological information and knowledge through the financial support of universities and research institutes. Also, pay attention to the industrial orientation of the research conducted by universities and research institutes.
- Provide incentive and support the effective establishment of proximity between firms oriented towards fruitful interactive learning and systemic innovation processes. Actions of this kind can be made through providing catalogs of related firms with related and complementary activities, making trade literature available, organizing fairs, attracting skilled talents and encouraging co-operative programs.

6.3.3 Timely Assessing the Inertia Governance Infrastructure

In this study, the inter-city comparison between Shenzhen and Dongguan, China indicates that governance differences indeed shape the innovation pattern of the respective regional innovation systems. Therefore, there is a need for monitoring and assessing the performance of governance infrastructure, and more importantly, adjusting the governance modality to meet the changes in the market, industrial organization and technology.

According to the analysis of the globalized grassroots production system in Dongguan, it is important to escape from the inertia caused by the competence trap and vested interests if evolution towards an interactive innovation system is to be achieved. In this respect, there is a call for the formation of network governance to unfasten the vested interests among the local developmental agencies, in which external agencies that have different interests from the local ones should be involved in generating incentives, developing technological alternatives and nurturing the required systems. Specifically speaking, a strategic plan of industrial development can be carried out with the identification and assessment of emerging changes in technological regimes, technological opportunities and patterns of demand that push the market into new technological areas (Smith 2000). There is clearly policy failure in the identification process due to the limited competency of public players to follow industrial dynamics. As a result, network governance should actively involve firms and industrial associations. Furthermore, dialogue between the public and private institutions should be encouraged. In summary, assessing the inertia of governance infrastructure is an important part of systems-oriented policies.

References

Arrow KJ (1962) The economic implications of learning by doing. Rev Econ Stud 29(80):155–173
Asheim BT, Coenen L (2005) Knowledge bases and regional innovation systems: comparing Nordic clusters. Res Policy 34(8):1173–1190
Braczyk HJ, Cooke P, Heidenreich M (1998) Regional innovation systems: the role of governances in a globalized world. Routledge, London
Chesbrough HW, Teece DJ (1996) When is virtual virtuous. Harv Bus Rev 74(1):65–73
Cohen WM, Levinthal DA (1990) Absorptive capacity: a new perspective on learning and innovation. Adm Sci Q 35(1):128–152
Cooke P, Heidenreich M, Braczyk HJ. (eds) (2004) Regional innovation systems: the role of governances in a globalized world. Routledge, London
Dollar D (2005) Globalization, poverty, and inequality since 1980. World Bank Res Obs 20(2):145–175
Humphrey J, Schmitz H (2008) Inter-firm relationships in global value chains: trends in chain governance and their policy implications. Int J Technol Learn Innov Dev 1(3):258–282
Meyer S (2011) Informal modes of governance in customer producer relations: the electronics industry in the greater Pearl River Delta (China). Franz Steiner Verlag, Leibzig
Nelson RR, Winter SG (1982) An evolutionary theory of economic change. Harvard University Press, Bonston
Parhankangas A, Arenius P (2003) From a corporate venture to an independent company: a base for a taxonomy for corporate spin-off firms. Res Policy 32(3):463–481
Smith K (2000) Innovation as a systemic phenomenon: rethinking the role of policy. Enterp Innov Manag Stud 1(1):73–102
Yeung HWC (2009) Regional development and the competitive dynamics of global production networks: an East Asian perspective. Reg Stud 43(3):325–351

Chapter 7
Appendices

Abstract The appendix chapter provides supplementary and illustrative information. The first section is the formulation of the 2009 PRD firm questionnaire, upon which the empirical investigation primarily relied. The second section illustrates the selecting procedures for the clustering analysis used in Chap. 3. The variable product technology used in the regression analysis of the study is based on International Standard Industrial Classification of all Economic Activities, which is displayed in the third section. In the last section, the development of Shenzhen electronics industry in the 1980s and 1990s is demonstrated with all key big events described in details.

7.1 Appendix A: Firm Questionnaire

Fact Sheet

1. Please name your most important product in terms of sales (e.g. notebooks, DVD player).

   _____

2. What percentage of sales did your company generate with its most important product category in 2008? _____%

3. In what year did your company start its operation in the e? _____

4. With which product did your company start its operation in the PRD (e.g. notebooks, DVD player)

   _____

5. What share of sales did your company generate with markets in the following regions in 2008?

   Sum of shares = 100 %
   ___ % Chinese mainland

___% HK
___% Taiwan
___% Rest of the World

6. **Please indicate share of your company's sales in 2008 according to the following categories in domestic (D) and international (I) respectively.**

D	I	
___%	___%	**Original equipment manufacturer (OEM):** products manufactured by your company according to design specifications provided by buyers or parent company
___%	___%	**Original design manufacturer (ODM):** products developed and designed by your company according to performance requirements of buyers or parent company
___%	___%	**Original brand manufacturer (OBM):** products developed and designed by your company and sold under own brand

7. **How is your company registered in the PRD?**

    ☐ State-owned ☐ Collectively-owned ☐ Private
    ☐ Wholly foreign-owned enterprise (incl. HK, MA, TW)
    ☐ Chinese-foreign equity joint venture (incl. HK, MA, TW)
    ☐ Chinese-foreign cooperative joint venture (incl. HK, MA, TW)

8. **If foreign owned (incl. HK, MA, TW): where does the main foreign investment come from?**

    _____

9. **If privately owned: Has your company been founded as a private company or has it been privatized in the past?**

    ☐ Privately founded ☐ Privatized

10. **Does your company belong to an enterprise group?**

    ☐ Private group ☐ state-owned group
    ☐ does not belong to an enterprise group

11. **Is your firm located in an industrial park?**

    ☐ Yes ☐ No

    If <u>yes</u>, please name the industrial park: _____

12. **Please give information about the supplier of your core parts and components (high-tech inputs such as CPU).**

    a) **Where is this supplier located?**

    _____ (city/province/country)

    b) **What kind of firm is this supplier?**

    ☐ Foreign customer ☐ Domestic customer
    ☐ Other foreign firms in the same industry

7.1 Appendix A: Firm Questionnaire

☐ Other domestic firms in the same industry

c) **How long has your company been working with this supplier?** _____ years

13. **Where does your company perform the following activities?** (Multiple answers possible)

	Location of activities
Production	
Management	
Finance	
Sales/Marketing	
R & D	
Training activities	

14. **Are the heads of the following units of your firm members of the main owner's family or the Chinese Communist Party (CCP)?** Member of …

Owners' Family	CCP	
☐	☐	CEO
☐	☐	Finance department
☐	☐	Production department
☐	☐	Marketing/ Sales department
☐	☐	Technical department
☐	☐	Human resource department

15. **Is there an official office of the Chinese Communist party (CCP) in your firm?**

    ☐ Yes   ☐ No

    If **yes**, how many persons are working in that office? _____

16. **Does your company has work union?** ☐ Yes   ☐ No

17. **What is the educational background of your workers in management and technical activities?**

Management staff	Technical staff	
___%	___%	Senior High or below
___%	___%	Vocational degree
___%	___%	Bachelor degree
___%	___%	Master degree
___%	___%	PhD

18. **How many sales in RMB has your company realized in 2008?** _____ millions of Yuan

19. **What is your company's average annual growth rate?**

    Sales

2007: _____% First half 2009: _____%

Net Profit

2007: _____% First half 2009: _____%

A. Market and Strategy

1. **This Unit of your company can be described as:**

    ☐ Individual enterprise
    ☐ Headquarter of a multi-firm company
    ☐ Regional Headquarter
    ☐ Affiliated of multi-firm company

2. **Which statement is most suitable to describe the strategic orientation of your firm? Your firm....**

    ☐ is oriented towards business opportunities in established markets
    ☐ just responds to incoming orders
    ☐ focus on upgrading its capabilities and position in the value chain
    ☐ follows emerging trends
    ☐ is introducing new brands or products to set new market trends
    ☐ tries to enter specialized markets with low degree of competition

3. **Which sources of finance did your firm use 2008?** (Share of financial sources in %, 100% in total) **Please, assess the accessibility of these sources** (1 - not accessible, 5 - easy accessible)

    ____% Chinese Bank   1 2 3 4 5
    ____% HK Bank   1 2 3 4 5
    ____% Foreign Bank   1 2 3 4 5
    ____% Stock market   1 2 3 4 5
    ____% Parent/ Affiliated company 1 2 3 4 5
    ____% Family members & friends 1 2 3 4 5

4. **Does your working capital primarily rely on bank loans?** ☐ Yes   ☐ No

5. **Does your investment capital primarily rely on bank loans?** ☐ Yes   ☐ No

6. **Does your firm feel the pressure of upgrading from following factors?** (1 - not important, 5 - very important)

    The rising cost of production   1 2 3 4 5
    Market competition   1 2 3 4 5
    The reducing number of foreign orders 1 2 3 4 5
    Government policies   1 2 3 4 5

    The Pearl River Delta (PRD) region, including Hong Kong and Macao, will be built into *"a globally competitive"* and *"vigorous area in Asia Pacific"* by 2020, says the National Development and Reform Commission (NDRC).

7.1 Appendix A: Firm Questionnaire

7. **How does the outline of the NDRC affect your business strategy?** (Multiple answers possible)

    ☐ New Investments in production
    ☐ Development of own brands
    ☐ More Research and Development
    ☐ No Change of business strategy
    ☐ I don't know the outline/plan of the NDRC

8. **Does your firm have special activities to improve the business activities?**
    ☐ Yes ☐ No

    If **yes**, please specify the activities. (Multiple answers possible)

    ☐ Work organization   ☐ Training programs
    ☐ Supply Chain Management   ☐ R&D
    ☐ Brand development   ☐ other Marketing activities

9. **In terms of upgrading, which statement is most suitable for your company? Upgrading is...**

    ☐ to stay in this business and increase the value added significantly.
    ☐ to diversify products and production.
    ☐ to switch completely to different products.

B. Organization and Marketing

10. **How many persons on average have been employed in your firm?**

    2007 ____ first half of 2009 on average ____

11. **How many persons on average have been employed in your firm in the following departments?**

2007	first half of 2009 on average
____	____ Production workers
____	____ Technical staff
____	____ Marketing/Sales
____	____ Management staff

Outsourcing is subcontracting a process, such as product design, manufacturing or other business functions to a third-party company. Insourcing is the opposite; it is defined as the delegation of operations within a business to an internal (but 'stand-alone') entity that specializes in that operation.

12. **a) Did your firm has any new business activities since 2007?** ☐ Yes → continue ☐ No → Jump to Q.13

12. **b) if yes, which are new activities?**

    ☐ Marketing/Sales (market research, consumer advertisement)
    ☐ Finance (Accounting/Bookkeeping)
    ☐ Production

☐ Research and Development
☐ Human Resources (Training/Recruitment)

12. c) Where does the new activity come from?

_____ (City/Province/Country)

12. d) What is the relationship between you and the firm that gave you this activity?

☐ Parent company or affiliated companies
☐ Foreign customers
☐ Foreign firms within the same sector
☐ Domestic customers
☐ Domestic firms within the same sector

13. a) Did your firm give any activities to other companies since 2007?

☐ Yes → continue ☐ No → Jumpt to Q.14

13. b) if yes, which are new activities?

☐ Marketing/Sales (market research, consumer advertisement)
☐ Finance (Accounting/Bookkeeping)
☐ Production
☐ Research and Development
☐ Human Resources (Training/Recruitment)

13 c) Where does the new activity come from?

_____ (City/Province/Country)

13. d) What is the relationship between you and the firm that gave you this activity?

☐ Parent company or affiliated companies
☐ Foreign customers
☐ Foreign firms within the same sector
☐ Domestic customers
☐ Domestic firms within the same sector

*Relocations activities cover the shift of a business unit or selective departments **within** the firm to a new location.*

14. How important are the following factors for your relocation activities?
(1 - not important, 5 - very important)

If no relocation, please jump to the next question.

Lower production costs	1 2 3 4 5
Availability of skilled workers	1 2 3 4 5
Better infrastructure	1 2 3 4 5
Preferential Policy	1 2 3 4 5

7.1 Appendix A: Firm Questionnaire

15. **Does your firm undertake marketing activities?**

    ☐ Yes → continue   ☐ No → Jumpt to Q.17

16. **a) Since when does your firm performs the following marketing activities?**

    Branding   _____
    Market Research _____
    Advertising   _____

16. **b) If your firm perform marketing activities: What are the goals of these?**
    (Multiple answers possible)

    ☐ Increase or maintain market share
    ☐ Enter new markets
    ☐ Increase visibility or exposure for products
    ☐ Response to Government Incentives for branding
    ☐ Increase the ability to adapt different client demands
    ☐ Develop stronger relationships with customers

17. **If your firm doesn't perform marketing activities: What are the reasons?**
    (Multiple answers possible)

    ☐ No need for marketing activities
    ☐ Outsourced
    ☐ Lack of qualified personal
    ☐ Parent Company is doing marketing activities
    ☐ Lack of funds within the Enterprise
    ☐ Lack of finance from sources outside the company (venture capital, public sources of funding)

C. Product & Process Development

18. **Since when does your firm design products by yourself?** _____
    (If not any, please leave it blank)
    Since when does your firm develop products by yourself? _____
    (If not any, please leave it blank)

19. **How important are the following activities for your company in the last 3 years?** (1 - not important, 5 - very important)

Introduce whole set producing lines	1 2 3 4 5
Self installing producing lines	1 2 3 4 5
Process engineering	1 2 3 4 5
Reverse engineering	1 2 3 4 5
Industrial design	1 2 3 4 5
Research and development	1 2 3 4 5

20. **a) How old are your machines and equipment on average?**

    ☐ earlier than 1980   ☐ 1980s   ☐ 1990s
    ☐ 2000–2005   ☐ newer than 2005

b) **Please roughly indicate the expenditure-to-sales ratio in upgrading machines and equipment in ....**

2007 __% first half 2009 __%

21. **Has your company set up technological center?** ☐ Yes ☐ No

    If **yes**, what kind of level:
    ☐ state-level TDC   ☐ province-level TDC
    ☐ municipal-level TDC   ☐ others

22. **Please tick the following patent that your company own.** (*multiple choice*)
    If not own any patens, please leave blank.

    ☐ Invention patent   ☐ Utility model patent
    ☐ Design patent   ☐ International patent

23. **Please evaluate the change of the following performance in your company in the past 3 years due to technological and innovation inputs.** (1-no change, 5-significantly change)

Cost reduction	1 2 3 4 5
Better product quality	1 2 3 4 5
More product function	1 2 3 4 5
More attractive product design	1 2 3 4 5
More flexible Production	1 2 3 4 5
New product category upgrading (e.g. from mp3 to mp4)	1 2 3 4 5

24. **Does your firm introduce any new products in the past 3 years?**

    ☐ Yes → continue   ☐ No → Jumpt to Q.17

25. **Please roughly indicate the expenditure-to-sales ratio in product development in....**

    2007 __% first half 2009 __%

26. **What share of total sales in the past 3 years is realized with new or significantly improved products? ...** _____%

27. **How important are the following ways in new product idea generation for your firms?** (1 - not important, 5 - very important)

1) Own idea generation and development	1 2 3 4 5
2) Reverse Engineering	1 2 3 4 5
3) Purchase product licenses	1 2 3 4 5
4) Orders from Parent company	1 2 3 4 5
5) Orders from Foreign customers	1 2 3 4 5
6) Orders from Domestic customers	1 2 3 4 5
7) Market report of sales agent	1 2 3 4 5
8) Market report of universities or research institutes	1 2 3 4 5

7.1 Appendix A: Firm Questionnaire

28. **How important are the following ways in acquiring necessary equipment or software in the process of new product development and production?**
(1 - not important, 5 - very important)

1) Acquisition from parent company	1 2 3 4 5
2) Acquisition from foreign customers	1 2 3 4 5
3) Acquisition from domestic customers	1 2 3 4 5
4) Others	1 2 3 4 5

29. **How important are the following ways in acquiring technical experience and know-how in the process of new product development and production?**
(1-not important, 5-very important)

1) Engineers sent by parent company	1 2 3 4 5
2) Engineers sent by foreign customer	1 2 3 4 5
3) Engineers sent by domestic customer	1 2 3 4 5
4) Engineers sent to foreign lead firms or customers	1 2 3 4 5
5) Engineers sent to domestic lead firms or customers	1 2 3 4 5
6) Engineers sent to universities	1 2 3 4 5

30. **How important are the following ways in getting into contact with your partners for new product development?** (1-not important, 5-very important)

1) Active searching (e.g. exhibitions, internet)	1 2 3 4 5
2) Business contacts (e.g. recommendation from partners)	1 2 3 4 5
3) Personal contacts (recommendation from family members and friends)	1 2 3 4 5

31. **a) Does your company obtain technological upgrading related knowledge from universities or research institutes?**

    ☐ no jump to Q.32   ☐ yes, name of UNI/RI:_____

    b) Please assess the importance of the following criteria considered for selecting universities as sources. (1 - not important, 5 - very important)

Reputation	1 2 3 4 5
Expertise/Research quality	1 2 3 4 5
Propensity for industrial co-operation	1 2 3 4 5
Uni is in the same city as your company	1 2 3 4 5
Uni is in Guangdong	1 2 3 4 5
Uni is in other provinces in China	1 2 3 4 5
Personal relationships	1 2 3 4 5

    c) Please mark the channels obtain knowledge and technologies from universities or research institutes.

Informal exchange	1 2 3 4 5

Technical services/testing	1 2 3 4 5
Technical consulting	1 2 3 4 5
Managerial consulting	1 2 3 4 5
Joint research/publication/patenting	1 2 3 4 5
Licensing of univ. technology	1 2 3 4 5
Buying of univ. technology	1 2 3 4 5
Staff mobility/training/internships	1 2 3 4 5

D. Human Resources

Recruitment practices

32. **Does your firm choose the following channels to recruit technical (T) and managerial (M) staff?** (Multiple answers possible)

	T	M
Vacancy advertisement in newspaper		
Company website, job-listing websites		
Government affiliated agency		
List of cold callers		
Personal networking/Recommendation		
Job fairs		

33. **When you employ technical (T) and managerial (M) staff, which candidate will you hire?**

    (one answer per column)   T   M
    A candidate with at least 3 years' experience working experience ☐   ☐
    A less experienced, but cheaper candidate   ☐   ☐

34. **On which contractual basis do you hire technical (T) and managerial (M) staff?** (Multiple answers possible)

	a) Without contract	b) Short-term contract	c) Long-term contract
T	____ %	____ %	____ %
M	____ %	____ %	____ %

    Short term indicates less than one year

35. **For which staff does your company use a recruitment agency?** (multiple answers possible)

    ☐ temporary   ☐ fixed term   ☐ permanent

36. **If you hire staff temporarily and/or fixed term, why?** (multiple answers possible)

☐ To adjust the size of the workforce.   ☐ To avoid Labour Law regulation.
☐ The company does not hire anybody permanently.
☐ Temporary and/or fixed term staff costs less.

Professional development and skills training

37. **To whom does your company offer job skills training and/or professional development?** (Multiple answers possible)

    ☐ No staff → Jump to Q43
    ☐ Management staff   ☐ Technical staff   ☐ Other staff

38. **When does the staff receive training and/or professional development?** (Only one answer)

	T	M
One time (e. g. when entering the company)	☐	☐
More often, but on an irregular basis.	☐	☐
On a regular basis	___days/year	___days/year

39. **How much did your company spend on training and/or professional development in…?**

    2007:___Yuan   first half 2009:___Yuan

40. **How important are the following ways to organize training?** (1 - not important, 5 - very important)

In-house training	1 2 3 4 5
Training through related agents or agencies	1 2 3 4 5
Training through parent company or affiliated company	1 2 3 4 5
Training through foreign customers	1 2 3 4 5
Training through domestic customers	1 2 3 4 5
Training through suppliers	1 2 3 4 5
Training through universities and research institutes	1 2 3 4 5

41. **Does your firm use the following ways to organize the trainning** (multiple answers possible)

    ☐ Coaching or mentoring
    ☐ Instructor-led workshops or courses
    ☐ Online tutorials and guided programs
    ☐ On-the-job training
    ☐ Printed materials (manuals, booklets)

42. **How important are the following aims in training the technical staff?** (1-not important, 5-very important)

Learn to use the equipment	1 2 3 4 5
Learn to maintain the equipment	1 2 3 4 5
Learn to repair the equipment	1 2 3 4 5
Learn to improve the equipment	1 2 3 4 5

Learn to develop new products/equipments   1 2 3 4 5

Staff retention

43. **How often do the following problems occur with your highly qualified staff?**
 (1 - never, 5 - very often)

High wage claims	1 2 3 4 5
Staff leaving after training	1 2 3 4 5
High absenteeism rate	1 2 3 4 5
Lack of practical skills	1 2 3 4 5
Insufficient quantitiy	1 2 3 4 5

44. **How does your company prevent highly qualified staff from leaving?** (<u>Multiple answers possible</u>)

    ☐ paid leave   ☐ materials (e.g. cellphone)
    ☐ housing   ☐ personal professional development
    ☐ health or accident insurance   ☐ higher salaries
    ☐ holidays or occasions

E. **Concluding Part:** *Personal Networks*

45. **How important are personal network to public officials for fulfilling the following tasks?** (1 - not important, 5 - very important)

Access to technology	1 2 3 4 5
Access to bank loans	1 2 3 4 5
Access to government funds	1 2 3 4 5
Access to reliable policy information	1 2 3 4 5
Recruitment of skilled personal	1 2 3 4 5
Access to export license	1 2 3 4 5
Access to domestic market license	1 2 3 4 5

46. **How often per week do you or key personal of your firm have formal contact with representatives of the government?**

    ___ times per week with representatives of the Local People's Congress or the local People's Political Consultative Conference
    ___ times per week with representatives of the Provincial/ National People's Congress or the Provincial/ National People's Political Consultative Conference
    ___ times per week with representatives of communist party

47. **Did your firm participate in the following supporting policy programs in the last 3 years?**

    If <u>**yes**</u>, please specify the yearly value of received funds in Yuan or if your firm received tax incentives.

7.1 Appendix A: Firm Questionnaire

Supporting Policies	Yuan	Tax incentives
High-tech enterprise identification program		Y/N
Innovation or upgrading funds from the local government		Y/N
Innovation or upgrading funds from the provincial government		Y/N
IPR advantage firms nurturing project		Y/N
Difficult firms certification and subsidies		Y/N

48. **What is the educational background of the CEO/ Managing director?** (Multiple answers possible)

    ☐ Bachelor degree
    ☐ Master degree
    ☐ Doctor degree
    ☐ Overseas Study experience
    ☐ Attended courses at Central Party School (CPS)
    ☐ None of the Above

49. **What is the work experience of the CEO/Managing director?** (Multiple answers possible)

    ☐ Worked in a private-owned company
    ☐ Worked in a state-owned company
    ☐ Worked as a government official
    ☐ Overseas work experience
    ☐ None of the above

Scale Standards:

1) Accessibility

1	2	3	4	5
Not access.	A little access.	Normal access.	Easily access.	Very easily access.

2) Importance

1	2	3	4	5
Not important	A little important	Of normal importance	Important	Very important

3) Significance

1	2	3	4	5
Not significant	A little significant	Of normal significance	Significant	Very significant

4) Frequency

1	2	3	4	5
Never	Seldom	Sometimes	Often	Very often

7.2 Appendix B: Test of Clusterin Solution

The choice of the clustering number is determined by two fitness criteria in the statistical sense: their Bayesian Information Criteria (BIC) and the Akaike Information Criteria (AIC). Most importantly, the interpretability of the model should be taken account of in order to ensure the theoretical soundness.

In Table 7.1, it can be concluded that the 3-cluster solution fits best according to BIC criteria, while the 4-cluster solution fits best according to the AIC criteria. In the latent class model, the BIC criteria decide the number of clustering in a more conservative way than the AIC criteria. In this way, the interpretability should be applied to make a choice for the mixed pattern.

In Table 7.2, I show the 4-cluster solution. In this solution, it is possible to identify the intensive interactive learning cluster (cluster 1) and the weak interactive learning cluster (cluster 4). However, cluster 2 and cluster 3 are quite similar in the scope and intensity of interactive learning and therefore do not differ from each other in a significant way. In order to derive a parsimonious and well interpreted result, I finally used the 3-cluster solution as the basis for the empirical analysis.

Table 7.1 Selection criteria by class

Classes	BIC	AIC
2	13198.8	12868.7
3	13075.1	12578.1
4	13137.5	12473.4

7.3 Appendix C: Classifying Product Technology

Table 7.2 The 4-cluster solution

		Probability of high evaluation[a]	Cluster 1	Cluster 2	Cluster 3	Cluster 4
Origins of innovation ideas	Own idea collection	0.81	0.61	0.47	0.44
	Reverse engineering	0.82	0.56	0.48	0.37
	Licensing	0.60	0.28	0.21	0.08
	Demand from parent company	0.54	0.32	0.20	0.06
	Demand from foreign customers	0.88	0.69	0.46	0.19
	Demand from domestic customers	0.91	0.63	0.50	0.41
	Market reports of sales agents	0.70	0.40	0.39	0.13
	Market reports of universities or research institutes	0.50	0.13	0.18	0.02
Support of equipment and software	Support from parent company	0.38	0.18	0.08	0.02
	Support from foreign customers	0.90	0.58	0.25	0.03
	Support from domestic customers	0.94	0.42	0.42	0.41
	Own purchase	0.29	0.14	0.06	0.22
Support of related technical know-how and experience	Engineers sent by parent company	0.34	0.17	0.03	0.03
	Engineers sent by foreign customers	0.87	0.42	0.23	0.00
	Engineers sent by domestic customers	0.85	0.41	0.39	0.16
	Engineers sent to foreign lead firms or customers	0.85	0.44	0.33	0.05
	Engineers sent to domestic lead firms or customers	0.81	0.49	0.53	0.29
	Engineers sent to universities	0.55	0.18	0.32	0.05
Interacting mode in the innovation process	Active searching	0.95	0.90	0.71	0.63
	Business contacts	0.98	0.81	0.66	0.55
	Personal contacts	0.71	0.27	0.35	0.25
Share of each cluster (%)		17	28	25	30

[a] Probabilities that the firm in each cluster give a high evaluation, i.e. important (4) or very important (5)

7.3 Appendix C: Classifying Product Technology

Note: This table is made jointly with the project member Daniel Schiller based on International Standard Industrial Classification of all Economic Activities, Rev 3. Products not included in the table are classified as low tech.

Category	Description
High	Internal combustion piston engines for aircraft, and parts thereof
High	Reaction engines
High	Turbopropellers
High	Parts for turbojets or turbopropellers
High	Power-generating machinery
High	Automatic data-processing machines; magnetic or optical readers, machines for transcribing data and machines for processing such data
High	Electrical apparatus for line telephony or line telegraphy
High	Transmission apparatus for radio-telephony, radio-telegraphy, radio-broadcasting or television
High	Telecommunications equipment, n.e.s.
High	Parts and accessories suitable for use solely or principally with the apparatus of division 76
High	Electrodiagnostic apparatus for medical, surgical, dental or veterinary purposes, and radiological apparatus
High	Diodes, transistors and similar semiconductor devices; photosensitive semiconductor devices (including photovoltaic cells); light-emitting diodes
High	Electronic integrated circuits and microassemblies
High	Piezoelectric crystals, mounted; parts, n.e.s., of the electronic components of group 776
High	Optical instruments and apparatus, n.e.s.
High	Compasses; other navigational instruments and appliances; surveying, hydrographic, oceanographic, hydrological, meteorological or geophysical instruments and appliances; rangefinders
High	Instruments and apparatus for physical or chemical analysis (e.g., polarimeters, spectrometers)
High	Oscilloscopes, spectrum analyzers and other instruments for measuring or checking electrical quantities instruments and apparatus for measuring or detecting alpha, beta, gamma, X-ray, cosmic or other ion radiations
High	Arms and ammunition
High	Hearing-aids (excluding parts and accessories)
High	Pacemakers for stimulating heart muscles (excluding parts and accessories)
Medium-high	Steam turbines and other vapour turbines, and parts thereof, n.e.s.
Medium-high	Internal combustion piston engines for propelling vehicles
Medium-high	Internal combustion piston engines, marine propulsion
Medium-high	Parts, n.e.s, for the internal combustion piston engines of subgroups 713.2, 713.3 and 713.8
Medium-high	Other gas turbines
Medium-high	Parts for the gas turbines of heading 714.89
Medium-high	Rotating electric plant, and parts thereof, n.e.s.

7.3 Appendix C: Classifying Product Technology

Category	Description
Medium-high	Gas generators, distilling or rectifying plant, heat-exchange units and machinery for liquefying air or other gases
Medium-high	Agricultural machinery (excluding tractors), and parts thereof
Medium-high	Wheeled tractors
Medium-high	Coal or rock cutters and tunnelling machinery
Medium-high	Other boring or sinking machinery
Medium-high	Coal or rock cutters and tunnelling machinery, not self-propelled
Medium-high	Other boring or sinking machinery, not self-propelled
Medium-high	Scrapers, not self-propelled
Medium-high	Parts for boring or sinking machinery of heading 723.37 or 723.44
Medium-high	Textile and leather machinery, and parts thereof, n.e.s.
Medium-high	Paper and pulp mill machinery, paper-cutting machines and other machinery for the manufacture of paper articles; parts thereof
Medium-high	Printing and bookbinding machinery, and parts thereof
Medium-high	Food-processing machines (excluding domestic); parts thereof
Medium-high	Other machinery and equipment specialized for particular industries; parts thereof, n.e.s.
Medium-high	Machine tools working by removing metal or other material
Medium-high	Machine tools for working metal, sintered metal carbides or cermets, without removing material
Medium-high	Medical, surgical or laboratory sterilizers
Medium-high	Driers for agricultural products
Medium-high	Driers for wood, paper pulp, paper or paperboard
Medium-high	Driers, n.e.s.
Medium-high	Vacuum pumps
Medium-high	Compressors of a kind used in refrigerating equipment
Medium-high	Air compressors mounted on a wheeled chassis for towing
Medium-high	Centrifuges (including centrifugal driers), n.e.s.
Medium-high	Filtering or purifying machinery and apparatus, for liquids or gases
Medium-high	Parts for the pumps, compressors, fans and hoods
Medium-high	Parts of the machines and apparatus
Medium-high	Ball- or roller bearings
Medium-high	Taps, cocks, valves and similar appliances for pipes, boiler shells, tanks, vats or the like
Medium-high	Gears and gearing; ball screws; gearboxes and other speed changers
Medium-high	Clutches and shaft couplings (including universal joints)
Medium-high	Calculating machines; accounting machines, ticket-issuing machines, incorporating a calculating device; cash registers
Medium-high	Photocopying apparatus incorporating an optical system or of the contact type, and thermocopying apparatus
Medium-high	Other office machines (e.g., addressing machines, automatic banknote dispensers)

Category	Description
Medium-high	Television receivers
Medium-high	Turntables (record-decks) and record-players
Medium-high	Sound-recording and other sound-reproducing apparatus; video-recording or reproducing apparatus
Medium-high	Microphones and stands therefor; loudspeakers; headphones, earphones and combined microphone/speaker sets; audio-frequency electric amplifiers; electric sound amplifier sets
Medium-high	Boards, panels (including numerical control panels), consoles for electrical control or the distribution of electricity
Medium-high	Optical fibre cables
Medium-high	Electrical insulators of ceramics
Medium-high	Television picture tubes, cathode-ray (including video monitor cathode-ray tubes)
Medium-high	Other electronic valves and tubes (including television camera tubes)
Medium-high	Batteries and electric accumulators, and parts thereof
Medium-high	Electric filament or discharge lamps; arc lamps; parts thereof
Medium-high	Electromechanical tools for working in the hand, with self-contained electric motor; parts thereof
Medium-high	Electrical capacitors, fixed, variable or adjustable (pre-set)
Medium-high	Electromagnets; permanent magnets
Medium-high	Electrical signalling, safety or traffic control equipment
Medium-high	Electric sound or visual signalling apparatus (e.g., bells, sirens, burglar and fire-alarms)
Medium-high	Carbon electrodes, carbon brushes, lamp carbons, battery carbons and other carbon articles used for electrical purposes
Medium-high	Motor cars and other motor vehicles principally designed for the transport of persons
Medium-high	Motor vehicles for the transport of goods and special-purpose motor vehicles
Medium-high	Road motor vehicles, n.e.s.
Medium-high	Parts and accessories of motor vehicles
Medium-high	Motor cycles (including mopeds) and cycles, motorized and non-motorized; invalid carriages
Medium-high	Trailers and semi-trailers; other vehicles, not mechanically-propelled; specially designed and equipped transport containers
Medium-high	Railway vehicles (including hover trains) and associated equipment
Medium-high	Lamps and lighting fittings (including searchlights and spotlights), n.e.s.
Medium-high	Illuminated signs, illuminated name-plates and the like
Medium-high	Parts of the portable electric lamps (excluding storage batteries)
Medium-high	Instruments and appliances, n.e.s., for medical, surgical, dental or veterinary purposes

Category	Description
Medium-high	Meters and counters, n.e.s.
Medium-high	Instruments and apparatus for measuring or checking the flow, level, pressure or other variables of liquids or gases
Medium-high	Measuring, controlling and scientific instruments, n.e.s.
Medium-high	Automatic regulating or controlling instruments and apparatus
Medium-high	Parts and accessories for machines, appliances, instruments and apparatus, n.e.s.
Medium-high	Photographic (other than cinematographic) cameras
Medium-high	Photographic flashlight apparatus
Medium-high	Cinematographic cameras and projectors; parts and accessories thereof
Medium-high	Microfilm, microfiche or other microform readers
Medium-high	Image projectors, n.e.s.
Medium-high	Photographic (other than cinematographic) enlargers and reducers
Medium-high	Spectacle lenses of other materials

7.4 Appendix D: Development of Shenzhen Electronics Industry in 1980s and 1990s

1979	In March, *Guangdong oversees Chinese farm management Bureau* signed with Hong Kong Ganghua Electronics Corporation in Beijing to establish Guangming oversees Chineses electricity firm that undertakes product processing
	In July, *Guangdong Electronics Industry Bureau* decided to establish out-oriented processing base in Shenzhen Special Zone and three state-operated factories in North Guangdong were relocated to Shenzhen
	In September, *Guangdong Planning Committee* approved the establishment of Guangdong Electronics Assembly Plant in Shenzhen (now as the Huaqiang Electronics Industry Company) which subordinated to the leadership of *Guangdong Electronics Industry Bureau*
	Later, *Communication Army Division in the General Staff Headquarter* invested and established Hongling Electric Appliance Processing Plant (now as Shenzhen Electric Appliance Corporation) in Shenzhen
	In December, *Fourth Ministry of Machine Building* transferred a group of technical backbone staff in Guangzhou 750 Factory to Shenzhen in order to establish Shenzhen Electronics Assembly Plant (now as Shenzhen Aihua Electronics Corporation). Meanwhile, *Third Ministry of Machine Building* established Shenzhen Office of China Aeronautical Technology Important Company (now as Avic Shenzhen Company)

	Later, China Foreign Investment Management Committee approved that *Oversees Chinese Town Economic Development Parent Company* joint venture with Hong Kong Ganggua Electronic Corporation, establishing the first industrial joint venture in Shenzhen "Guangdong Guangming oversees Chineses Electronics Industry Company" (now as Shenzhen Konka Group Company Ltd.) that produces recorders and televisions
1980	In April, Shenzhen Revolutionary Committee approved the establishment of joint venture Xinhua Electronics Plant which *Shenzhen Industry Bureau* provided land and Hong Kong Xinyou Trade Corporation provided capital and equipment
	In May, *China Electronics Technology Important & Export Company (Shenzhen Division)* was established
1981	In January, *Guangzhou Electronics Industry Bureau* decided to establish Guangdong South-China Radio Factory in Shenzhen based on 8571 Factory
	In March, *Guangdong Electronics Industry Bureau* signed the joint venture contract with China Electronic Device Parent Company in Beijing to establish Shenzhen Huaqiang Electronic Device Company (now as Shenzhen Yuehua Electronic Device Company)
	In September, *Shenzhen Industry Bureau* established Shenzhen Kangle Electronics Corporation jointly with Haerbing Fourth Radio Factory
	In October, *Shenzhen Industry Bureau*, China Zhenhua Electronics Company (earlier as the 837 Factory) and Hong Kong Luks Group Co. LTD. jointly established Shenzhen Huafa Electronics Corporation and introduced the 14"–22" production line of color TV with annual output of 100,000
	Later, *Fourth Ministry of Machine Building* (now as Zhenhua Electronics Group) invested in Shenzhen and established Shenzhen Huayun Electronics Co., Shenzhen Shenyun Electronics Co. and Shenzhen Huafa Electronics Co.
1982	In January, Shenzhen Electronics Industry Company (earlier under the same system with Shezhen Industry Bureau) was established, which was in charge of municipal electronics companies and new self-invested electronics companies
	In July, wholly Hong Kong invested company "Hong Kong Luks Industry Corporation was approved by Shenzhen City Government. Later on, it introduced 14"—18" production line of color TV with annual output of 100,000
	In December, Guangdong Guangming oversees Chineses Electronics Industry Company (now as Shenzhen Konka Group Company Ltd.) introduced 4"—20" production line of color TV with annual output of 300,000
1983	In April, wholly Japan invested Sanyo Electric Machinery Corporation was registered
	In June, *state-owned and operated 8571 factory* jointly established Shenzhen Yuebao Electronics Joint Company with Baoan Industrial and Transportation Bureau under the approval of Guangdong Province Economic Development Committee
	In Octorber, 70% of the color TV produced by Shenzhen Huafa Electronics Corporation after introducing the production line was exported and the output was expanded to annually 450,000
	In November, China Aeronautical Technology Important Company (now as Avic Shenzhen Company) and Beijing Computer Parent Company jointly invested and established China's first LCD and LCD model design and manufacturing specialized company "Shenzhen Tianma Micro-electronics Corporation". In 1984, the first TN-LCD production line went into operation in Tianma Corporation

7.4 Appendix D: Development of Shenzhen Electronics Industry ...

	In December, *Shenzhen Electronics Industry Development Coordination Committee* was established jointly by related management departments of state ministry-related companies, Guangdong Electronics Industry Bureau and Shenzhen City Government
1984	In January, Shenzhen Office of Ministry of Electronics was approved and established in order to manage the ministry-related companies and public institutions in Shenzhen
	Later, Guangdong Guangming oversees Chineses Electronics Industry Company (now as Shenzhen Konka Group Company Ltd.) introduced new color TV production line and began the production of 14"color TV
	In January, *Guangdong Electronics Industry Bureau* transferred a group of technical staff that produces audio-head in Guangdong South-China Electronics Company to Shenzhen and jointly invested with Baoan Industrial Development Company to establish Yuebao Electronics Joint Company
	In April, Shenzhen Huaqiang Electronics Industry Company and Japan Sanyo Electric Device Co. Ltd. established the joint venture Huaqiang-Sanyo Electronics Co. Ltd with the contract period of 15 years
	In May, Shenzhen Electronics Industry Parent Company (now as Saige Group) and China Electronic Device Parent Company established the joint venture "Color Kinescope Company". In July, Shenzhen development project of color kinescope was established to take charge of the plan, application and negotiation of joint ventures
	In July, Shenzhen Division of Avic Technology, Industrial and Trade Company and Southern Aerodyne Mechinery Company established Shenzhen Shennan Electric Circuit Corporation. In 1985, double and multi-layer printed circuit production line was introduced from USA and was then able to produce printed circuits no more than six layers
	In August, China Shenzhen Color TV Parent Company was jointly invested and established by Ministry of Electronics Industry, China Electronic Device Parent Company and Shenzhen City
	In October, Shenzhen Futian Industrial Development Company and Sixth Experimental Factory of Post and Telecommunications Institutes of Changchun jointly invested 900,000 Yuan to establish Shenzhen Changhong communication equipment company limited, which undertakes the research and production of small-volume exchange equipment
	In December, Shenzhen City Government approved the establishment of Shenzhen Xianke Laser Television (SAST) Co. Ltd., which includes Laser Sender Company, Laser CD Company and Laser Technology Research Institute. The total investment was 300 million Yuan
	In December, China Academy of Sciences established China Kejian Co. Ltd. in Shenzhen
1985	In January, Esopn invested 10 million Hong Kong dollar and established wholly owned Yexin Technology (Shenzhen) Co. Ltd. It introduced the ESOPN printer production line with annual output of 300,000 and the export rate is 100%. In 1987, this production line went into operation
	In February, Shenzhen Zhongxing Semiconductor Co. Ltd was established, which investment reached 2.8 million Yuan and came jointly from aerospace system 691 Factory, Great Wall Industrial Company (Shenzhen) and Hong Kong Yunxing Electronic Trade Company

	In April, Shenzhen Xianke Laser Television (SAST) Co. Ltd. signed a technology cooperation contract on Laser singing and sight system with Holland Phillip Company
	In July, Technology Development (Shekou) Co. Ltd. invested 2 million dollar introduced American hard disc magnet head production line with annual output of 1 million
	In September, Shenzhen Electronics Group Company (later as Saige Group) was established under the approval of Shenzhen City Government, which unifies 117 companies among 178 companies on voluntary basis. It was then one of the four experimental sites of enterprise group of electronics industry in China
1986	In March, Shenzhen Electronics Industry Association was established based on *Shenzhen Electronics Industry Development Coordination Committee*
	Wholly American Owned Company Flextronics was established in Shenzhen and a mainframe production line with annual output of 1 million was introduced
1987	In May, China Computer Development Company relocated the production base to the South and established China Great Wall Computer Development Company (Shenzhen) and starts the production of Great Wall series of PC
	In July, Shenzhen Computer Industrial Association was established
	In July, Japanese owned company "Topresearch Circuit Board Co. Ltd." was established in Shenzhen and mainly produced double and multi-layer circuit boards
	In September, Modern Electronics Industry Co. Ltd. (also called MAC, co-established by China Electronic Device Company and Hong Kong Kangmao Development Co. Ltd.) was established with the investment of 128 million dollar and purchased three production line of big-screen color kinescope and technology from American General Electric. The annual production was 1.5 million
	In September, Shenzhen Huawei Technology Co. Ltd was established and was one of the earliest private-owned technological enterprises at that time
	In October, Shenzhen Electronics Products Quality Control Center was opened
	In November, Shenzhen Sanda Electronic Industry Company was established based on Shenzhen Office of Ministry of Electronics
	Hong Kong invested company "Nantai Electronics (Shenzhen) Co. Ltd." was established in Shenzhen with the operation of processing with supply material
1988	In March, The first specialized electronic parts supply market in China "Saige Electronics Supply Market" was opened which is build by Shenzhen Saige Group
	Later on, Shenzhen Zhongxing Semiconductor Co. Ltd cooperated with Beijing College of Post and Telecommunication to jointly research the company's first generation digital customer exchange machine ZX500
	In June, China Tongguan Telecommunication Co. Ltd. jointly invested 13.5 million dollar with Canada Northern Telecommunication Co. Ltd. and established joint venture Tongguang-Northern Co. Ltd. Meanwhile, Meridian ISDN digital customer exchange machine production line was introduced with an annual output of 100,000
	In July, Shenzhen Telecommunication Industry Co. Ltd. established cooperative enterprise "Shenzhen Guangtong Development Co. Ltd." Meanwhile, 8.5 million dollar was invested and a leading optical fiber production line was introduced with an annual output of 25,000 chip km
	In August, Shenzhen Software Industrial Association was established

1989	In January, Shenzhen Saige Group and Janpa Hitachi established joint venture "Shenzhen Saige-Hitachi color Monitor Co. Ltd.", and produced the leading 21" color kinescope with an annual output of 1.6 million
	In August, China Electronics & Information Industry Group, Shenzhen Saige Group and Hong Kong Kangmao Development Co. Ltd established the joint venture "Shenzhen Zhongkang Glass Co. Ltd." and introduced the technology and equipment from America and Japan with an annual output of 4.3 million sets
	In September, Shenzhen Lenovo Computer Co. Ltd was established with an annual output of 1 million computer mainframe
1992	In March, Shenzhen Konka Electronics Stock Limited Corporation and Shenzhen Huafa Electronics Stock Limited Corporation went into stock market
	Shenzhen Foxconn Precision Parts Plant was established
	Skyworth Group was established with the headquarter in Hong Kong and produced color TV, VCD, DVD and satellite receiver
	In December, Shenkou Development and Technology Co. Ltd and American Conner Company established the joint venture "Shenzhen Kangnuo External Equipment Co. Ltd." which was the first in China that produces Hard Disk Driver with an annual output of 2 million HDD. All the products were then exported
	In December, Shenzhen Zhongxing-Weixiantong Equipment Co. Ltd. was established under the support of shareholder units of aerospace system. A proportion of technical backbone staff and managerial backbone staff in Shenzhen Zhongxing Semiconductor Co. Ltd. were transferred
1993	In March, Shenzhen Zhongxing New Telecommunication Equipment Co. Ltd. was established which was jointly invested by 691 factory under aerospace system, Shenzhen Guangyu (Group) Company and Zhongxing-Weixingtong Equipment Co. Ltd. The company first applied in China the operation system of "state-own and private-operated"
	Shenzhen Jindie Software Technology Co. Ltd. was established
1997	Shenzhen City Government conducted assets reorganization between Shenzhen Liming Electronics Industry Co. Ltd., Shenzhen Guangtong Development Co. Ltd. and Xingsuo Optical Cable Telecommunication Industry Company (the last two subordinated under Shenzhen Telecommunication Industry Co. Ltd.) and put them under centralized management of Shenzhen Special Zone Development Group Co. Ltd. Meanwhile, Shenzhen City Government established Shenzhen Tefa-Liming Photoelectric Co. Ltd. and Shenzhen Tefa Information Stock Limited Company, where the latter one was divided into Shenzhen Tefa Optical Fiber Co. Ltd. and Shenzhen Tefa-Xingsuo Optical Cable Telecommunication Industry Company
2003	In January, Shenzhen Electronic Chamber of Commerce was established
	Shenzhen City Government decided that auto electronics is the main orientation for development in the process of industry structural adjustment
	In July, Shenzhen Electronic Chamber of Commerce signed the "Memo of Cooperation" with Hong Kong Electronic Chamber of Commerce

Source: SECC 2004

The manufacturer's authorised representative in the EU is Springer Nature Customer Service Centre GmbH, Europaplatz 3, 69115 Heidelberg, Germany. If you have any concerns regarding our products, please contact ProductSafety@springernature.com

Printed and bound by CPI Group (UK) Ltd, Croydon, CR0 4YY

23/03/2026

02076379-0005